I0522235

The
Generative
Age

The Generative Age

Artificial Intelligence and the Future of Education

Alana Winnick

ConnectEDD Publishing
Hanover, Pennsylvania

Copyright © 2023 by Alana Winnick

All rights reserved. No part of this publication may be reproduced, distributed, or transmitted in any form or by any means, including photocopying, recording, or other electronic or mechanical methods, without the prior written permission of the publisher, except in the case of brief quotations embodied in critical reviews and certain other noncommercial uses permitted by copyright law. For permission requests, contact the publisher at: info@connecteddpublishing.com

This publication is available at discount pricing when purchased in quantity for educational purposes, promotions, or fundraisers. For inquiries and details, contact the publisher at: info@connecteddpublishing.com

Published by ConnectEDD Publishing LLC
Hanover, PA
www.connecteddpublishing.com

Cover Design: Kheila Casas

The Generative Age: Artificial Intelligence and the Future of Education by Alana Winnick. —1st ed.
Paperback ISBN 979-8-9874184-6-8

Praise for *The Generative Age*

This book is truly a guide for those who are on the fence about AI in Education, as I was. It provides concrete examples of lesson plan ideas and provides a framework for those who are looking to bring it to the masses in their school community. This is a must read for all educators who are even a little curious about bringing AI into their students' lives!

—Jay Strumwasser | Director of Technology

The Generative Age: Artificial Intelligence and the Future of Education is a must-read in this era of educational transformation. Seamlessly blending theory with practice provides realistic solutions to challenges educators face with AI implementation. Its practical and ethical considerations make it an invaluable guidebook for any educational leader navigating the AI landscape. Undoubtedly, this book is the torchlight in the foggy realm of AI in education.

—Joseph Durney, Ed.D. | CEO, School District Leader, Professor

This book is a comprehensive look at Generative Artificial Intelligence. Alana offers an engaging way to explore and comprehend what AI is and how we can harness its power. This book will help us all better understand and use artificial intelligence to support our students.

—Laurie Guyon | Coordinator for Model Schools

Alana did a fantastic job in streamlining and addressing the current issues involved with artificial intelligence for educational leaders, presenting them in a well-organized, accessible format. She shares personal stories that are practical and illustrate the complex realities of the evolving technology and how they can be harnessed to benefit learning.

—Tim Needles | Educator and Author of *STEAM Power*

In this guidebook Alana offers an accessible way to understand and dive into the world of Generative AI. She gives concrete examples that educators at all levels can relate to and implement into any classroom. I see myself sharing this with my colleagues and learning and growing with this book as a model.

—Meredith Dutra | Instructional Technology Specialist

As we enter the generative age when it comes to Artificial Intelligence in schools, leaders will need high quality guidance and inspiration to integrate AI with fidelity. Alana's book provides not only that guidance but real-world applications of AI that will empower teachers and leaders as we enter this next phase of learning with technology. A must read!

—Carl Hooker | co-Founder K12Leaders

In a time when everyone has questions about Artificial Intelligence and its impact on our educational institutions, this book has the answers. Whether you work with students, or are leading a staff through this AI moment, you'll find a clear path to follow and case studies to reference in *The Generative Age: Artificial Intelligence and the Future of Education*!

—A.J. Juliani | Best-Selling Author and Founder of *Adaptable Learning*

The Generative Age: Artificial Intelligence and the Future of Education is a as timely as it is essential for all educators. With her visionary mindset and commitment to educational transformation, Alana presents practical strategies and actionable insights that empower both teachers and leaders to navigate the complex landscape of AI integration into our work. This book serves as a compass, guiding all of us to make both incremental adjustments and bold, transformational changes to ensure that we and our students are well-equipped to tackle the challenges of the future.

—Dr. Adam Pease | Assistant Superintendent for Curriculum & Instruction

For educators in primary, secondary, or higher education, *The Generative Age: Artificial Intelligence and the Future of Education* by Alana Winnick is a must read. This is one of those few books that will be relevant for at least the next decade considering how quickly the AI landscape is changing. The book addresses the key aspects of artificial intelligence (AI) particularly in the areas of fostering AI culture, navigating academic integrity, and implementing effective teaching strategies in the AI age.

This book is an excellent resource that combines theoretical perspectives with practical advice. It will help educators gain a comprehensive understanding of the opportunities and challenges presented by AI no matter where they fall on the AI spectrum.

—Dr. Daren Khairule | Director of Academic Technology

As AI becomes increasingly prevalent, this AI Guidebook arrives at just the right moment. Alana has demystified AI and made it accessible to teachers by demonstrating multiple, diverse ways of making our lives easier! For example, this book helped make my assessments more creative by easily generating text in French. Not only did AI create my French reading and listening comprehension texts, but it also created multiple choice and open-ended questions for those texts. What a time saver!

—Katell McNamara | French Teacher

The Generative Age: Artificial Intelligence and the Future of Education by Alana Winnick is an essential read for educators who are curious about the role of AI within the profession. Right from the beginning, Alana provides a reassuring message, emphasizing that AI's purpose is not to replace teachers. Instead, she skillfully illustrates how it is a powerful tool that can enhance instructional practices and streamline tasks. The book offers revolutionizing perspectives on familiar and popular frameworks that are already innovating learning in today's classrooms and suggests insightful ways on leveraging AI to personalize instruction so that all learners are successful. *The Generative Age: Artificial Intelligence and the Future of Education* invites readers to discover the transformative potential that AI offers educators who are guiding their learners through the Generative Age.

—Marina Lombardo | Elementary Educator

Artificial intelligence (AI) has taken the world of education by storm. With previous disruptive technologies, schools often took years to mindfully consider adoption strategies, but AI continues to evolve on an almost daily basis. This book serves as an outstanding primer for anyone looking to learn more about how this technology stands to reimagine teaching and learning experiences in our schools. But more than that, it contains impactful use cases and stories that provide the reader with ready-made takeaways designed to help you put AI to use for your class or school immediately!

—Jesse Lubinsky | Education Evangelist, Adobe Inc.

Dedication

Zachary, Harrison, Charlie, Piper, and Hailey,

You are the future! Always remember that you have the power to become whoever you want to be and achieve anything you set your mind to. You are capable of extraordinary things beyond your wildest imaginations. Pursue your passions, embrace the opportunities that come your way, and never be afraid to dream big! The world is truly yours for the taking, and I am here to support you, believe in you, and watch in awe as you create a better tomorrow.

This book is a testament to the incredible potential that lies within each one of you. It is a reminder that the future of education goes beyond what we teach and how we teach it, but about the extraordinary individuals like you who will learn, evolve, and become the architects of a world that we can only begin to imagine.

Stay curious, be true and kind to yourself, and most importantly, never stop dreaming! I love you.

Table of Contents

CHAPTER 1

A Universe Unveiled

Our Journey into The Generative Age

Welcome to The Generative Age, a time of unprecedented growth and innovation. The world as we knew it (past tense) transformed on November 30th, 2022, the infamous day that ChatGPT by OpenAI was released. This generative artificial intelligence (AI) software generates human-like text within seconds (hence the word "generative") and has disrupted the educational system. As educators, it is our mission to prepare students to become the architects of a world that we can only begin to imagine. This is your guidebook to shaping the future of learning in your school or classroom by harnessing the power of this disruptive technology. We will embark on this exciting journey together, embrace the possibilities, and challenge the status quo as we redefine education for future generations. The time for action is now—let's pave the way to a brighter, (AI)-empowered future for all.

If you're reading this, I'm sure you've been striving to support your teachers/students in a post-pandemic world that seems to be shift-ing faster than ever; I can relate to that. With the emergence of gen-erative AI, we can't go back to the way things used to be. The role of technology in education has become crucial and we must stay at the forefront of these advancements in order to provide our students with the skills necessary to thrive in an increasingly complex world. Among the many technologies that have the potential to transform education, generative AI is arguably the most promising and disruptive technology of our time.

> Generative AI is arguably the most promising and disruptive technology of our time.

In this book, we'll dive into the potential of generative AI to reshape education, exploring both its opportunities and its challenges. Together, we will examine the responsibilities that we must embrace to ensure that this powerful technology is integrated effectively and ethically into our schools and communities. The following chapters will provide invaluable insights, strategies, and best practices that will empower you to make informed decisions as you prepare your schools and classrooms for the AI-driven world of tomorrow.

This book is more than just an exploration of AI's impact on education; it is also a call to action. As educators and leaders, you are uniquely positioned to shape the future of learning and ensure that AI is harnessed responsibly. I hope that by the time you reach the end of this book, you will feel inspired, well-informed, and ready to initiate transformative change within your schools.

Prepare for launch. T-minus 3... 2... 1... Liftoff!!

My Journey into The Generative Age

My journey began by stumbling upon generative AI, not really sure what it could do. Like I do with every other new technology, I play! I used it all the time and pushed it to its limits, figuring out what it can do really well and where it falls short. At first, I mostly used it for emails, but then I started asking it to help me with pretty much every single thing I do, both personally and professionally; from generating show notes for a podcast, to helping me craft the perfect response to a sensitive message or email–it could really do it all.

One day in January 2023, Dr. Cameron Fadjo, Assistant Superintendent at Pleasantville Union Free School District, visited our district to learn about our online registration system, and while he was there, we began discussing ChatGPT, because, well, who wasn't talking about it at that time?! We recognized early on that there was going to be a need to educate teachers and leaders about the various aspects of this emergent technology. We also realized that, as the educational landscape undergoes a massive transformation, educators will need actionable steps to embrace generative artificial intelligence.

I embarked on a journey that I never imagined would lead me where I am today: producer, host, and editor of *The Generative Age*, a podcast powered by The New York State Associa-tion for Computers and Technologies in Education (NYSCATE), and author of this very book you're holding in your hands. Cameron and I developed "The Big Five" in the first podcast episode as a way to bring actionable steps to our audience. These still stand true today, so if you take anything from this book, please remember this:

The Big Five:

1. Set expectations with teachers and students around generative AI
2. Build awareness with teachers and students around generative AI
3. Focus on critical thinking and reasoning
4. Use generative AI to augment your professional work
5. Revisit the assessments you are asking of your students (focus on the process, not the product)

Scan here for The Generative Age podcast

As a struggling creative writer and speller, you can only imagine how challenging this experience has been for me. I cannot imagine how students with dyslexia or any other writing difficulties must feel when faced with similar challenges.

Overwhelmed and unsure of where to begin, I turned to my trusted companion, artificial intelligence, for guidance. The synergy between my own human intelligence and artificial intelligence enabled me to push the boundaries of my imagination, generate new ideas, and go deeper into the ideas I already had, all while honoring my individual perspective and my distinct voice. I ran my book through an AI-detection tool, which we will cover later in the book, and passed with flying colors! With a commitment to maintaining academic and intellectual integrity, I used these tools for initial research, followed by incorporating proper citations from other, more reputable sources.

I also began using the dictation feature on my document, which allowed me to compose passages within minutes because I can speak much faster than I can write. Once I dictated my thoughts and turned them into text, I relied on artificial intelligence to refine the content; removing filler words, summarizing my ideas, expanding on different areas; this created coherent text much quicker than I ever could, which allowed me to write the first draft of this book in one month. Afterwards, I revised the passages to reflect my own voice and add personal touches, ensuring that it sounds just like me. In the sections that follow, you will see additional concrete examples of how you can use generative AI to create content in your own work.

As I reached the About the Author section of this book, I quickly realized that I needed a professional and flattering headshot. Being camera shy and not particularly photogenic, I procrastinated hiring a photographer until it was too late; I found myself in quite the predicament. What did I do? You guessed it—I turned to my reliable partner, artificial intelligence. By uploading 14 selfies to a website, an impressive array of 200 pictures were generated and I was able to successfully incorporate a glamorous headshot into my bio...take a look for yourself!

A Teacher's Journey into The Generative Age

Once upon a time, there was a dedicated and passionate teacher named Skeptical Ms. Sinclair. Ms. Sinclair has taught English and Creative Writing for over two decades and has seen the educational landscape evolve throughout her career, including the difficult transition to remote teaching during the pandemic. She loved her job and her students, but lately, she has been feeling worn out and frustrated, and the last thing she wanted to do was learn about yet another new technology that she wasn't comfortable with.

One day, her Director of Technology (DOT) presented at a faculty meeting and showcased a new generative AI tool that was just released. The educational system was changing way too fast, and she couldn't keep up. She had always been a traditional teacher, using tried-and-true methods that had worked for her in the past and always yielded positive outcomes for her students. She didn't see the need for all these new changes and new tools. Her DOT was excited about this new technology and encouraged all teachers to consider trying it. Despite the enthusiasm of her colleagues, Ms. Sinclair was skeptical about embracing this new tool. She was worried that it would replace her job, remove her authentic voice, and negate her decades of experience—of course she knew her students better than any computer program ever could. She was highly concerned that this new technology would have a negative impact on her students' creativity and repress their originality; she considered the use of this tool in writing to be cheating. Yet, she couldn't help but notice how enthusiastic her colleagues seemed to be about it—it was all anyone was talking about in the faculty lounge over lunch.

A Universe of (AI) Possibilities

Imagine a classroom that transcends the traditional boundaries of space and time, where learning is seamless, highly personalized, and limitless. Each student is immersed into a learning environment that is specifically designed to challenge and engage their unique needs and interests. Let's take a peek at some of the endless possibilities:

+ **Emily** wakes up to her AI assistant's friendly greeting. They review her agenda for the day, which ranges from school to extra-curricular activities. Over hot chocolate and a croissant, Emily practices her French with her AI-powered language tutor in a virtual café in Paris. Despite being in her kitchen, the virtual environment (or experience) is so realistic that she feels immersed in the culture, allowing her to hone in her language skills through context and conversation. Since she is not conversing with an actual human, she does not feel judged when her tutor corrects her annunciation. This allows her to push her vocabulary even further because she is not afraid to make mistakes.

- **Maya,** a second grader, is transported to an enchanting world filled with mythical creatures and magical landscapes. She embarks on an interactive creative writing adventure, interacting with the Alice-in-AI-Wonderland-esq beings. Maya weaves through the world, creating her story, and her AI tutor supports her exploration of language, creativity, and developing character backgrounds. The tutor provides feedback and suggestions to help Maya improve her creative writing skills, identifies areas of improvement, and provides targeted, on-the-spot exercises to help her develop her skills.

- **Winston** is a high school student who is passionate about political science. He is deeply engaged in a spirited debate with an AI-powered counterpart that is representing a different perspective on recycling. His AI opponent rebuts his statements and adds new perspectives, pushing his critical thinking and problem-solving skills, encouraging him to explore alternative viewpoints and develop a deeper understanding of the subject.

- **Julius** struggled to engage with traditional history lessons. He finds them boring and would often distract his peers by talking in class or passing notes. When his teacher introduced him to a new virtual environment where he could engage directly with historians, he discovered a newfound interest in the subject. He is now able to immerse himself in the historical event and interact with and learn directly from virtual historical figures, such as Leonardo Di Vinci, William Shakespeare, and Martin Luther

King Jr. This unique and immersive approach transformed Julius' understanding and appreciation of history, making learning a truly captivating experience.

- **Albert,** a middle school student, has always struggled with math and no one at home is capable of helping him with his complex homework. He had almost given up, and decided that "he's just not good at math, and that's OK" because he is great at many other things. When his teacher introduced him to his new AI-tutor, he started tackling challenging math problems, making quite a few mistakes at first. Instead of providing the correct answer, his AI-tutor nudged him to review his steps and apply the proper formula. It didn't correct him, instead it guided him, helped him identify where he went wrong, and how to apply the correct formula. It gave him the tools, information, and formulas he needed to understand both the problem and the process, allowing him to discover the solution on his own. This helped him build confidence in his math abilities and he no longer felt helpless when facing a difficult problem.

Does this seem futuristic? This world is much closer than you think, and the key to unlocking it lies within the integration of generative AI. These powerful learning experiences leverage other emerging technologies such as chatbots, voice and video communication, and virtual reality. In a TedTalk titled "How AI Could Save (Not Destroy) Education," (Khan, 2023) Sal Khan stated that if we put the right guardrails in place, we can use AI for the biggest positive transformation that education has ever seen. To accomplish this, every student will receive an artificially intelligent, but incredible, personal tutor, and every teacher will receive an artificially intelligent, but amazing teaching assistant.

Khan refers to Benjamin Bloom's 1984 2 Sigma Problem, which shows compelling data to prove that if each student were provided with a 1:1 tutor, then they could actually improve by two standard deviations (see diagram below)--yes, you heard that right, TWO standard deviations. He goes on to explain that with a 1:1 tutor, your average students can become exceptional students and your below average students can become above average students. The only issue stopping Bloom from scaling this model was cost, and generative AI disrupts that.

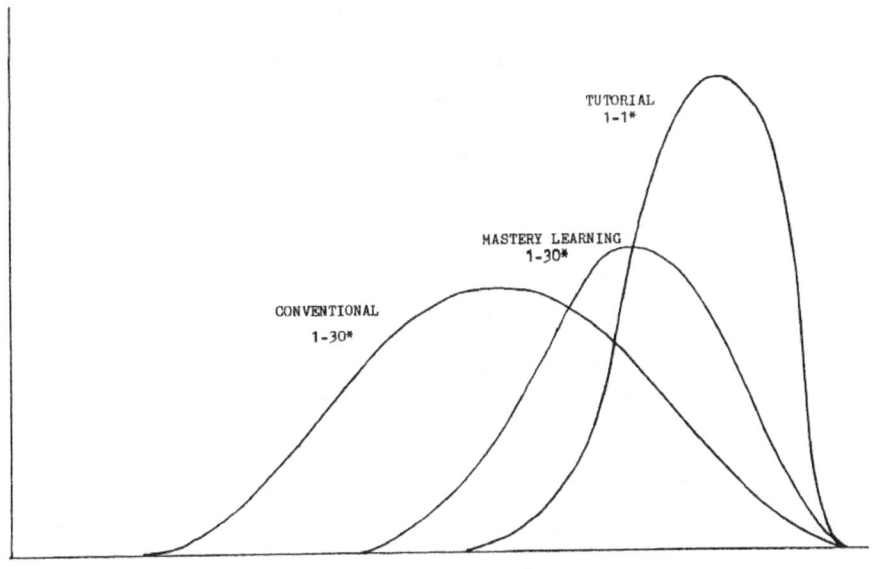

*Teacher-student ratio

(Bloom, 1984)

This new AI-powered world allows schools to become more like real-life and teachers to serve as mentors and facilitators, guiding their students through challenges and providing them with personalized support and encouragement. They are able to foster genuine connections with each student, ensuring that no one is left behind and empowering them to be the inspirational, supportive mentors they have always aspired to be. This is just the beginning, and the possibilities are limitless.

So, What Exactly is Generative AI?

To put it in simple educational terms, it's the world's smartest librarian. Imagine that your librarian has read and memorized every single book in the entire library and can recall all information at the drop of a hat and reformat it to any audience, from teaching fair use to kindergarteners to presenting the digital literacy curriculum to the board of education—and then rewrite it again within seconds based on your feedback. In simple terms, this is generative AI.

More technically speaking, generative AI is typically built using machine learning algorithms—Large Language Models (LLMs)—allowing them to understand, analyze, and generate (hence the word generative) human language in a coherent and con-textually relevant way. Generative AI can be multimodal, meaning it can analyze and output text, images, videos, and more. In this book, we will mostly discuss text-based generative AI, (I will be sure to also highlight other generative AI, such as art, design, and music), which can learn language patterns and structures by training on large text datasets, such as books, articles, and websites. During the training pro-cess, they analyze patterns in word sequences, enabling them to reason about grammar rules and contextual relationships. These tools predict and output a response by generating the next most probable word.

What makes this different from a search engine, such as Google, is that instead of redirecting you to existing sources on the internet, generative AI responds to your prompts with tailored information in a conversational tone. Humans analyze and make sense of the world around us, while AI is solely predicting and outputting what the next word should be based on probability, at least for now.

This technology is not new—when you write emails and it predicts what it thinks you will write next, that's generative AI. When you're on a website and notice a chatbot in the bottom-right corner, that's generative AI. Remember IBM's Watson (from a decade ago)? That's generative AI. It's been around for years, but this is the first time it is highly accessible and easier to use. These highly sophisticated models can generate coherent and contextually appropriate human-like text in a conversational tone within seconds and output everything from writing essays, creating poetry, answering questions, generating computer code, and providing 1:1 coaching on everything from reading comprehension to solving complex math problems, and beyond. Similar to how we rarely hear about Watson anymore, I would bet that we won't hear about ChatGPT soon either, because this technology will be integrated directly into other tools we use and love, from the Microsoft/Google suite to all your favorite educational technology applications.

CHAPTER 2

~

A World of (AI) Possibilities

Phase One: Time Well Spent

As you may have noticed in the previous chapter, I used the heading "A Universe of (AI) Possibilities," and here I refer to a *world* of possibilities. Just like a universe bursting with possibilities, the potential of AI in the classroom is vast and can feel a bit overwhelming. Since a universe is much larger than a world, we are going to start smaller, because you need to start somewhere. By focusing on a smaller scale (aka "a world") in phase one, we can create a manageable entry point that feels achievable. As you gain confidence, you can graduate into phase two which is more focused on student use as new instructional software is released. Remember, this is just the tip of the iceberg.

We often hear complaints from teachers regarding new initiatives, such as "I don't have enough time," "When am I supposed to do that?" and "Will I be compensated?" With most of their day spent in front of students, teachers struggle to allocate extra time to new initiatives or tasks, unless coverage can be arranged–which requires planning (time) on the teacher's part and an additional budget expense for administrators. Teachers in the United States have the highest burnout rate of all professions. More than 4 of every 10 teachers state that they feel burned out "always" or "very often" at work and more than 300,000 quit the profession between February 2020 and May 2022 (Smith, 2022).

According to the *U.S. Department of Education, Office of Educational Technology* (2023), on average, teachers report spending only 49% of their time with students; they spend 10.5 hours per week on preparation, 6.5 hours on evaluation and feedback, and 5 hours on administration (pg 31). We hope that educators want the best for their students, but we can't stretch or buy more time. When implementing new initiatives, administrators may even find themselves negotiating with unions; however, I have a solution that is cost-free, union-proof, and will grant teachers more time. Yes, you read that right: more time. That solution is generative AI. Imagine if teachers could decrease weekly preparation from 10.5 hours to six, which the report suggests, saving them 4.5 hours in that one category alone (pg 32). Like I said before, time is limited, so let's use it for things that matter, like connecting with other humans.

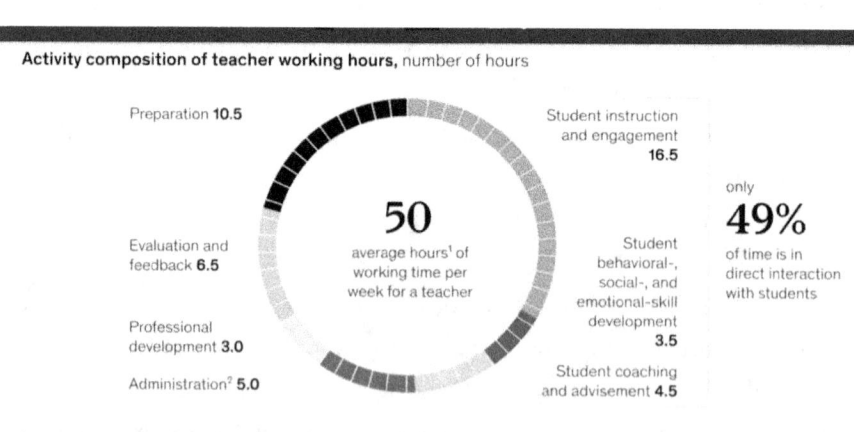

Activity composition of teacher working hours, number of hours

Preparation **10.5**

Student instruction and engagement **16.5**

only **49%** of time is in direct interaction with students

50 average hours[1] of working time per week for a teacher

Evaluation and feedback **6.5**

Student behavioral-, social-, and emotional-skill development **3.5**

Professional development **3.0**

Student coaching and advisement **4.5**

Administration[2] **5.0**

[1] Average for respondents in Canada, Singapore, United Kingdom, and United States.
[2] Includes a small "other" category.
Source: McKinsey Global Teacher and Student Survey

(U.S. Department of Education, Office of Educational Technology, 2023)

The technology we see today will not be a standalone software, it will be integrated into most of the technology we are already using, which is why I will not provide you with examples of specific tools or how to use them, because that will likely be outdated by the time this book is published. AI is evolving at

> **It will be a combination of teachers along with technology that will bring our students to the next level.**

a rapid pace, and we are just beginning to see the opportunities it provides us as educators. The important thing to keep in mind is that this technology will not replace teachers. It will be a combination of teachers along with technology that will bring our students to the next level.

Content Creation

Generative AI has the ability to produce a wide range of outputs, from intricate text compositions to creative works of art. Below are a few examples of its application:

Homework: Reflect, Create, and Accelerate

Writing:

1. Using a pencil and paper, write a 1,000 word reflection on what you've learned about generative AI. How long did it take you?

2. Using a computer, type a 1,000 word reflection on what you've learned about generative AI. How long did it take you?

3. Prompt ChatGPT, Bard, The New Bing, or any other generative AI tool to write a 1,000 word reflection using your original ideas. How long did it take you? (Prompt engineering framework provided in chapter 8).

Scan the QR code for tutorials.

Homework: Reflect, Create, and Accelerate

Art:

1. Color/paint an image that portrays your reflection. How long did it take you?

2. Search the internet for an image that accurately portrays your reflection. How long did it take you?

3. Prompt Canva, Adobe Firefly, Dall-E, or any other generative AI tool to create an image that accurately portrays your reflection. How long did it take you?

Scan the QR code for tutorials.

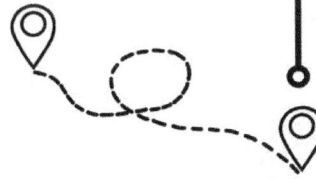

Homework: Reflect, Create, and Accelerate

Design:

1. Create a trifold poster on what you've learned so far. How long did it take you?

2. Create a PowerPoint/Google Slides with bullet points and images on what you've learned so far. How long did it take you?

3. Prompt gamma.app, slidesai.io, or any other generative AI tool to create a presentation that accurately portrays your reflection. How long did it take you?

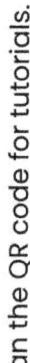

Scan the QR code for tutorials.

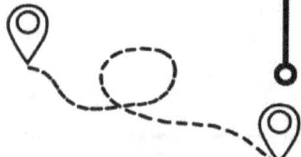

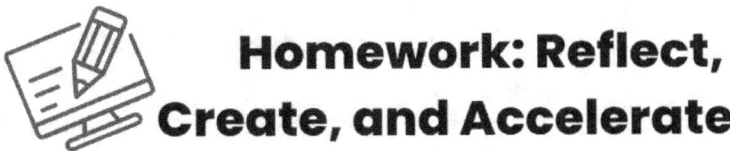

Homework: Reflect, Create, and Accelerate

Reflection Questions:

- What was the purpose of each assignment?

- Did the AI compromise the integrity of the assignment and the knowledge I was seeking from you?

- How much time did the AI-assisted tasks save you compared to the traditional tasks?

- Which method did you prefer for each subject?

Homework: Reflect, Create, and Accelerate

- How did the AI-generated art compare to your hand-drawn/painted image?

- Did the AI tool improve your efficiency in the design assignment?

- Are you ready to use these tools in your own practice?

- Has this assignment influenced your thoughts on AI's role?

- How will you repurpose the time that you just saved by incorporating AI?

Think of AI-use similarly to traveling by foot vs driving a car vs flying a plane—or better yet, a rocket ship. This technology saves a whole lot of time. Now, more importantly, what are you going to do with that time?

A Pressing Revolution: AI's Impact on Knowledge Dissemination

Let's hop in a time machine and travel back to the 1500's when the Gutenberg printing press was first invented. Before the printing press, books were very expensive and extremely difficult to produce, which meant that only the wealthy or those with connections to the church or government had access to them.

After the printing press, books became more widely available and affordable, a llowing e ven m ore p eople t o a ccess k nowledge a nd ideas. Both the printing press and generative AI revolutionized the production of written material, making it possible to generate text much more quickly and affordably than ever before, which has had a significant impact on the evolution of communication and knowledge dissemination.

What's interesting is that we are in that same transcendental moment as the invention of the printing press, which led to modern journalism and publishing. As AI continues to advance, it will lead to new forms of communication, learning, and collaboration. What do you think those will be?

The Human Element

I often hear the question: "Will AI take my job?" No, genera-tive AI will not take your job, in fact the U.S. Department of Education (2023) rejects this idea (pg 3), but someone who uses it might, because they will be much more efficient than someone who does not.

Generative AI can be a thought partner, but the thoughts should always start with the user. To harness its full potential, it is essential to maintain a balance between AI assistance and human intervention and view generative AI as a thought partner or supplementary resource rather than a complete solution. If you haven't done so already, try it out for yourself. Start by sharing your thoughts and asking the AI for suggestions. Go back and forth with your chatbot, but always make sure the final product is your own. For prompting support, refer to the Prompt Engineer Framework and tutorials in chapter 8. The AI tool can be a helpful resource, but it should not replace genuine thought and effort.

The *U.S. Department of Education, Office of Educational Technology* (2023), outlined seven recommendations for desired qualities of AI tools and systems in education (see image below). Notably, the first recommendation, *Humans in the Loop*, emphasizes the importance of human teachers being in control of critical instructional decisions, including the interpretation of data (pg 21).

Figure 14: Recommendation for desired qualities of AI tools and systems in education

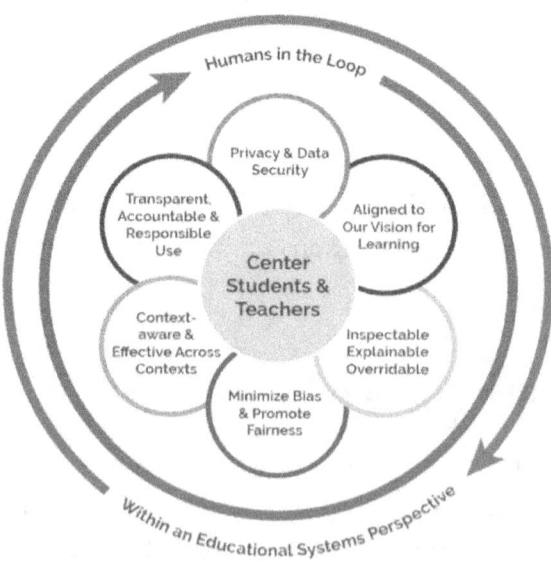

(U.S. Department of Education, Office of Educational Technology, 2023)

The Balancing Act

To achieve a balance between generative AI and human involvement, ask yourself:

1. What distinctive characteristics of my job should I preserve?
2. What aspects of my job are not essential to my unique qualities? Can AI assist me in performing first drafts of these tasks more efficiently?
3. When is it better to hand a task off to AI?

Transforming Administrative Work in The Generative Age

How can you use generative AI in your own work? You can probably use it for most things and it will eventually become your virtual assistant, integrated into everything you do–predicting what you will need even before you know you need it. Try it out. Push it to see what it can do better than you. Compare it to your own work–what do you think?

Why should you reinvent the wheel when you don't need to? You know what they say, work smarter, not harder–it will save you time. You can repurpose that time to focus on relationships that drive your organization. Examples of ways you can use generative AI in your own work today:

- **Agendas:** Create data driven agendas that address your specific needs and goals.
- **Flyers:** Create eye-catching and informative flyers for your events or initiatives.
- **Images:** Generate relevant and engaging images to enhance learning materials, presentations, and projects.

- **Emails:**
  - ○ **Informational emails to families and staff:** Quickly generate clear, concise, and effective emails. You can even ask AI to personalize your emails, but be mindful of inputting sensitive data if the tool does not adhere to federal and state data privacy and security regulations.
  - ○ **Replies:** Generate responses that address parent/teacher concerns and provide next steps for resolution. Generative AI can also help you anticipate potential objections or concerns and prepare responses in advance.
- **Letters of recommendation:** Streamline the letter-writing process and create compelling and personalized letters of recommendation for your teachers/students by inputting relevant information such as achievements, extracurricular activities, and personal qualities. Be careful about including students personally identifiable information (PII), which we will cover later in chapter 6.
- **Newsletters:** Generate newsletters that are informative, entertaining, visually appealing, and personalized with your class/school's news and events.
- **Policy creation:** Generate policies that are clear, concise, and aligned with your class/school's objectives. Input information about your class/school's policies, including specific requirements or restrictions. You can even ask the AI to suggest edits, additions, or deletions to ensure that the policy is comprehensive and easy to understand.
- **Presentations:** AI can help you write a clever title and description, then form a cohesive presentation by turning your ideas into slides, assisting with design layouts, speaker notes, images, and more.
- **(Board) Reports:** Efficiently and effectively communicate your class/school's performance by providing valuable insights into

your school's progress, achievements, and areas for improvement.

+ **Speeches:** Craft an impactful speech (commencement, retirement, etc.), adapting your message to your unique audience.

+ **Translation:** Translate content into multiple languages, enabling you to reach a diverse range of students/parents, which will foster inclusivity.

Although she didn't want to admit it, Ms. Sinclair was slowly starting to embrace the idea of generative AI and its potential to revolutionize her classroom based on the conversations she overheard in the faculty lounge. One weekend, she sat at her computer, curious, and determined to put AI to the test. She thought, "If this can help me save time, I'd have even more time to focus on my students."

In her first prompt, she asked the AI tool to generate an email to families, informing them about the upcoming Parent-Teacher Conference. Before her very eyes, the AI quickly generated a clear, concise, and engaging email within seconds! It would usually take her at least fifteen minutes to craft the perfect email including all the details. Feeling more confident in its capabilities, she asked the tool to write a letter of recommendation for one of her students who was applying for a summer writing program. She input the student's achievements, extracurricular activities, and personal qualities—leaving out all personally identifiable information (PII)—and the AI produced a compelling and personalized letter that showcased the student's talents—all within seconds. Ms. Sinclair was mesmerized!

Transforming Teaching in The Generative Age

Educators are able to create a more dynamic and personalized learning experience by combining AI-generated suggestions along with their own expertise and understanding of their students' unique needs. Below are examples of how generative AI can be used to aid teachers and enhance the learning experience for their students. Remember to remove students' personally identifiable information if the tool you're using is not compliant with your state's student data privacy and security laws.

- **Administrative processes:** Streamline administrative tasks, such as writing emails, IEPs, letters of recommendation, newsletters, report generation, and data analysis.
- **Content creation:** Generate engaging educational content, such as instructional videos, presentations, unplugged (screen-free) classroom activities, worksheets, quizzes, assessments, discussion questions, student exemplars, images, infographics, and supplementary materials.
- **Data-informed decisions:** Detect patterns in student data and generate data-informed teaching and learning recommendations. Be aware of bias and algorithmic discrimination, which we will cover later in the book.
- **Grading:**
  - Craft customized rubrics tailored to your specific learning objectives.
  - Streamline the grading process by assessing student work based on a rubric you created, providing consistent evaluations and feedback in a fraction of the time.
- **Improved accessibility:** Generative AI can break down language barriers and ensure educational content is accessible to a broader audience by providing real-time translations, leveled text, summarizations, explanations.
- **Lesson/unit planning:** Assist educators in designing lesson and unit plans by providing suggestions and ideas based on the subject matter and learning objectives. By inputting specific topics and goals, teachers can receive tailored starting points for their unit outlines and lesson plans, which they can modify according to their students' needs and school/department format.

+ **Reading/Listening comprehension passages (in any language):** Generate passages for any subject area and in any language, then rewrite the passage in multiple reading levels within seconds.
+ **Research assistance:** Aid with research tasks by summarizing articles, providing overviews of specific topics, and suggesting relevant sources. While AI should not replace comprehensive research, it can be a helpful starting point or time-saving tool. Be mindful that it may not be accurate, so it is essential to cross-check your information.
+ **Talking points:** Generate discussion prompts, debate topics, or lesson/presentation outlines.
+ **Video Review:** Create engaging and informative videos reviewing content from a unit/lesson.

Both curious and wanting the best education for her students, Ms. Sinclair decided to try out AI with instructional content. She experimented by prompting the tool to "generate ideas for creative writing prompts." Much to her surprise, AI provided her with unique and thought-provoking prompts, which would undoubtedly inspire her students and encourage them to think critically, empathetically, and explore diverse perspectives in their writing.

See below for writing prompts selected from a larger list generated by OpenAI (2023):

1. The world has been plunged into silence – no one can speak, sing, or make any noise. How do people communicate, and what unexpected consequences arise from this new reality?
2. Write a story from the perspective of an everyday object (e.g., a pencil, a shoe, a water bottle) as it experiences a day in the life of its owner.
3. A mysterious door appears in your bedroom that can transport you to any fictional world you choose. Describe your adventures and the challenges you face as you explore this new realm.

4. You discover a hidden map inside an old book at the library. The map leads you to a long-lost city filled with forgotten knowledge and ancient artifacts. Describe your journey and the secrets you uncover.

Ms. Sinclair's students loved the prompts and eagerly began crafting their own narratives, each bringing a different perspective to the table. As time went by, Ms. Sinclair noticed significant improvement in her students' enthusiasm, creativity, and writing skills. They eagerly shared their stories, offering constructive feedback to their peers, and engaging in meaningful discussions about their work.

Transforming Learning in The Generative Age

Generative AI has the potential to revolutionize learning by making it more engaging and interactive for students. AI allows us to tailor instruction to individual students, providing personalized feedback and adaptive content. Students can actively engage with generative AI tools to create artwork, compose music, or solve complex problems. Just like the calculator heightened the complexity of the math that we are capable of doing on our own, AI will heighten the complexity of the writing that our students are capable of producing–I am living proof of that with this very book. Outlined below are ways that students will be able to utilize generative AI in their own work–make sure that whatever tool you're using is compliant with your state's student data privacy and security laws:

- **1:1 Tutoring:** Engage in personalized sessions with AI-powered virtual tutors, providing additional support tailored to individual learning needs and preferences.
- **Breaking down tasks:** AI can help students break down tasks and assignments into manageable steps, providing personalized

guidance and support along the way. This can help students to be more productive, confident, and successful in their learning.

- **Images:** Generate relevant and engaging images to enhance presentations and projects.
- **Immersive learning experiences:** Provide hands-on, interactive experiences that deepen understanding and increase engagement in various subjects.
- **Improved accessibility:** Ensure educational content is accessible to all students by summarizing articles, leveling the text, translating, providing explanations for difficult topics, and more.
- **Language learning:** Assist students in their writing and reading comprehension skills. Students can translate documents and engage in interactive conversations with AI to practice grammar, vocabulary, and syntax.
- **Peer review and feedback:** Provide initial feedback on student work, identifying areas for improvement and offering suggestions for revision. This feedback should be considered supplementary to teacher-guided evaluation and peer review, as the AI may not always capture the intricacies of human language and thought.
- **Personalized learning:** Provide individualized feedback and guidance to students based on their unique learning needs and abilities. This can help identify improvement areas and tailor the learning experience to optimize student growth.
- **Presentations:** Students are able to form a cohesive presentation by easily turning their ideas into slides and assisting with design layouts and speaker notes.
- **Research assistance:** Aid with research tasks by summarizing articles, providing overviews of specific topics, and suggesting relevant sources. While AI should not replace comprehensive research, it can be a helpful starting point or time-saving tool. Be mindful that it may not be accurate, so it is essential for students to cross-check their information.

- ✦ **Support struggling writers:** Provide students with a starting point for their writing, which can alleviate the stress and anxiety that often accompanies the writing process for struggling writers, as well as increase their confidence in expressing their thoughts and ideas. This will allow students to focus on their ideas and the content of their work rather than getting bogged down in the mechanics of writing.
- ✦ **Talking points:** Generate debate topics, presentation outlines, or podcast scripts, which will stimulate critical thinking and enhance communication skills.

(If You) USEME-AI Model

The USEME-AI Model (Taylor, 2023) created by Stephen Taylor (Director of Innovation in Learning & Teaching at the Western Academy of Beijing) serves as a guiding framework for students as they embark on their journey into The Generative Age.

IF YOU USEME-AI ADAPTING TO AI IN SCHOOLS — QUESTIONS FOR STUDENTS

U	UNDERSTANDING THE TECHNOLOGY	Do I **UNDERSTAND** this AI tool, its purpose and how it can be used appropriately?
S	SUPPORTING MY LEARNING	How does this tool **SUPPORT** my learning and give me opportunities to think more deeply?
E	EXPECTATIONS FOR PURPOSEFUL LEARNING	Do I know the **EXPECTATIONS** for purposeful learning, academic integrity and safety?
M	MODELING POSITIVE INTERACTIONS	How am I **MODELING** good use of AI in interactions with my peers, through evaluating and discussing our work?
E	EVALUATING ETHICAL IMPLICATIONS	How am I considering the **ETHICS** of AI and the implications & opportunities of AI to make a difference?
-		
A	AUTHENTIC APPLICATIONS	How can AI enhance my work in solving **AUTHENTIC** & meaningful problems and making personal connections?
I	INSPIRING INNOVATION	How is my learning **INSPIRING** me to take meaningful action and try new things with innovation & creativity?

(If You) USEME-AI was created by **Stephen Taylor** For the Western Academy of Beijing

learn.wab.edu/innovation/ai
sjtylr.net/if-you-useme-ai

(used with permission from Stephen Taylor
and the Western Academy of Beijing)

As Ms. Sinclair continued to explore its capabilities, she discovered new ways to use AI. This time she put it in the hands of her students—of course she checked with her Director of Technology first and the tool she used was compliant with data privacy and security regulations (which can vary from state to state, but essentially refers to the requirement of vendors keeping student's data safe).

Ms. Sinclair considered the benefits of using AI for review and feedback. She knew that this would not replace her evaluation, but she believed that it could serve as a supplementary tool, offering initial feedback and suggestions for revision. Students could refine their work before sharing it with their peers or submitting it for grading. She decided that she would require students to submit a transcript of the chat along with their final assignment so that she could see HOW the tool was being used.

Language learning was another area where Ms. Sinclair saw potential for AI integration. She sat with one of her English Language Learners and showed her how to properly prompt the chatbot (I will teach you this in chapter 8) which guided her with corrections for grammar, vocabulary, and syntax, as well as how to translate documents to and from her native language.

Ms. Sinclair also recognized the potential for generative AI to support struggling learners and writers, especially one particular student, Struggling Sierra. She sat with Sierra and showed her how to input the assignment and ask the chatbot to break it down into manageable steps; it even provided her with a starting point for her writing. This tool would allow Sierra and other students to focus on their ideas and content, alleviating the stress and anxiety that they often feel over their writing assignments.

As you can see, generative AI holds tremendous potential for enhancing educational processes by streamlining content creation, providing research assistance, and offering personalized learning experiences. By understanding its capabilities and limitations, educators can save time and create a more engaging and meaningful learning environment for their students.

 # The History and Future of Research

What constitutes original research? Take a look at the following timeline; it highlights the significant shifts in research methods over the years, from relying primarily on physical resources like libraries and encyclopedias to embracing digital resources and generative AI:

1950's
- Libraries were the information source for research, including physical books, journals, and encyclopedias.
- Card catalogs were used to locate resources within libraries.

1960's
- Libraries continued to be the primary source of information for research.
- Photocopiers became commonplace, making it easier to reproduce and share information.
- Early computer databases were introduced, but they were not commonplace.

 # The History and Future of Research

1970's

- The introduction of electronic databases made searching for information more efficient and accessible.
- Early online library catalogs and interlibrary loan systems began to emerge, allowing users to access resources from multiple libraries.
- The first personal computers were introduced.

1980's

- Personal computers became more popular, and the use of electronic databases and online library catalogs grew.
- CD-ROMs emerged as a new way to store and distribute information.

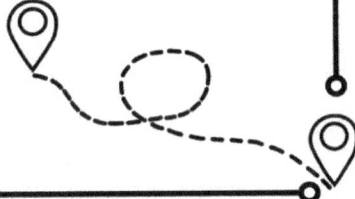

The History and Future of Research

1990's

- The World Wide Web revolutionized research, providing easy access to vast information.
- Search engines, such as AltaVista and Yahoo!, made it easier to find information online.
- Digital libraries, electronic journals, and databases became more prevalent, transforming how people accessed scholarly materials.
- Acceptable research increasingly included online resources, although print materials remained important.

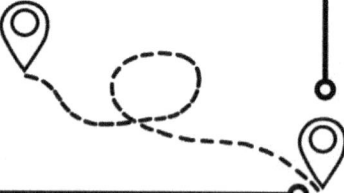

The History and Future of Research

2000's

- The dominance of Google as a search engine simplified and streamlined the process of finding information online.
- The rise of Wikipedia provided a comprehensive, user-generated encyclopedia that quickly became a popular research starting point (keywords: starting point - I will come back to this at the end of the timeline)
- Open access publishing gained momentum, allowing researchers to access and share scholarly materials freely.
- The use of print materials in research continued to decline; digital resources became the primary means of information access.

The History and Future of Research

2010's

- The growth of mobile devices and widespread internet access made information available anytime, anywhere.
- The rise of social media and online forums provided new platforms for sharing and discussing research and information.
- Machine learning and AI-driven search engines improved information retrieval and discovery.
- Acceptable research became predominantly digital, with online databases, e-journals, and websites serving as the primary sources of information.

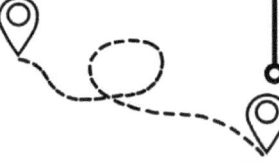

The History and Future of Research

2020's

- Generative AI provided a new way to generate content and synthesize information from diverse sources.
- The use of AI-driven tools and algorithms to analyze large datasets and extract insights became more commonplace.
- Acceptable research continued to evolve, encompassing AI-generated content, as long as they adhered to ethical guidelines and maintained academic integrity.

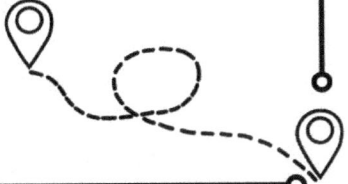

Wikipedia: Fact or Fiction?

Let's get back to Wikipedia. Is Wikipedia considered a reliable resource? When Wikipedia was first released, I struggled with this very question. Finally, I decided that it was "a great place to start." We don't know what we don't know, so it provides us with enough knowledge and context to learn what we are searching for. That is exactly how we should think of generative AI: every search could start here, but we should always question the AI-generated output. It is important to know that AI can provide inaccurate information, so you must identify a reliable source that confirms your findings—keep an eye out for that in the next chapter. As you can see, I used AI to help me generate research, but I then backed it up with legitimate citations from reliable sources. As research practices evolve, it's crucial to focus on academic integrity, ethical use, and critical thinking to ensure the quality and reliability of the information being used.

~

CHAPTER 3

Leading with Integrity
Fostering a Culture of AI Responsibility

Unraveling Ethical Challenges

As you learned in the previous chapter, generative AI can trans form education in numerous ways, from personalized learning to efficient administrative processes. As we embark on this journey to explore the world of generative AI in schools, we must remain mindful of the challenges and ethical dilemmas of integrating these tools into education. This section will guide us on addressing these concerns responsibly and thoughtfully.

To Ban or Not to Ban? That is the Question

At this point in the book, I hope I have convinced you of the value of harnessing the potential of this disruptive technology and how it can transform education and your school. If not, let's travel back in history. When cars were first invented, there were major concerns about their impact on the existing transportation system—yes, I am talking about horses and horse-drawn vehicles. Some cities and towns even banned cars. The introduction of cars represented a major technological and cultural shift, and it took time for society to adapt to the presence of cars and implement regulations to address these concerns. Can you imagine a present-day world with horse-drawn vehicles as our primary mode of transportation? I certainly cannot. I see this being the same with generative AI; the world simply needs time to accept and embrace it.

Many schools immediately blocked ChatGPT, some for academic dishonesty reasons, and others for data privacy and security concerns. The reality is, AI is having its "app store moment," (which refers to the iPhone's surge in popularity following the explosion of the app store) and this technology will inevitably be integrated into most educational technology tools that we already use and love, making it impossible to avoid.

The Moral Compass: Exploring Key Concerns

As we dive into the ethics of generative AI in education, several key considerations emerge that warrant attention and discussion:

+ **Academic integrity:** Using generative AI in student work can blur the lines between original and machine-generated content; this raises concerns about academic dishonesty and the potential decline of critical thinking and creativity.
+ **Accessibility and the digital divide:** As generative AI technologies become more prevalent in education, there is a risk that students without access to them may be left behind, increasing existing imbalances in educational opportunities.
+ **Algorithmic discrimination:** Generative AI models learn from the distribution of the data it is trained on. This can result in biased outputs if the training data is unrepresentative of certain

demographics or contains biased information. An example of this would be if the model was trained in the medical field and most of the Doctors were males and most nurses were females, that's exactly what the AI will assume. This can lead to unfair learning opportunities for certain populations, such as providing unhelpful hints or resources to a specific group of students or an early intervention recommendation based on gender and race… or perhaps in some instances it will be less biased than some humans?

- **Data privacy and security:** What information does it have access to? The use of generative AI requires the input of data, including the potential of students' personal information and academic records. Ensuring the privacy and security of this data and complying with data privacy and security regulations is paramount.

- **Ethics:** When is it unethical to pass important issues off to AI? When is it more responsible to pass important issues to AI due to human bias? Both of these scenarios are equally problematic.

- **Explainability:** While AI can recognize patterns and make accurate decisions, its limitation in providing an explanation poses a challenge. The data or evidence that was used to create AI-generated recommendations must be transparent to educators.

- **Incomplete data:** Incomplete data could lead to unexpected results, such as speeding up or slowing down a student's curricular pace, which could increase learning gaps.

- **Lack of contextual judgment:** AI systems analyze patterns and automate decisions, but they often lack contextual data and judgment, such as knowing that the teacher or student had a bad day.

- **Recursive self-improvement:** Aka "AI teaching AI." This involves AI using its own capabilities to enhance its performance. This has raised many concerns due to the fact that there is not a human directly controlling it. You may have heard this referred to as a "black box" because it is difficult for us to fully understand or predict how AI might evolve or make decisions over time.

Questioning the AI-generated Output

As AI tools continue to be integrated into education, we must instill responsibility and critical thinking in our teachers and students. They must understand that AI-generated content is not always accurate and may produce biased results, so it is important for everyone to question the output and make necessary adjustments. This concept is similar to media literacy, where students are taught to question the credibility of resources found online and identify reliable sources that confirm their findings. This also helps students take ownership of their work, ensure that it aligns with their unique perspectives and voices, and develop a sense of agency over the final product.

I recently watched a TedTalk featuring Yejin Choi, a computer scientist and Professor at the University of Washington, titled *"Why AI is incredibly smart—and shockingly stupid."* (Choi 2023). In the talk, she addresses the challenges and limitations of current artificial intelligence systems (of course this will change with time) and asks ChatGPT the following questions:

Prompt 1

(The top image depicts Choi's conversation with OpenAI.
The bottom image shows Alana's same OpenAI prompt with different results.)

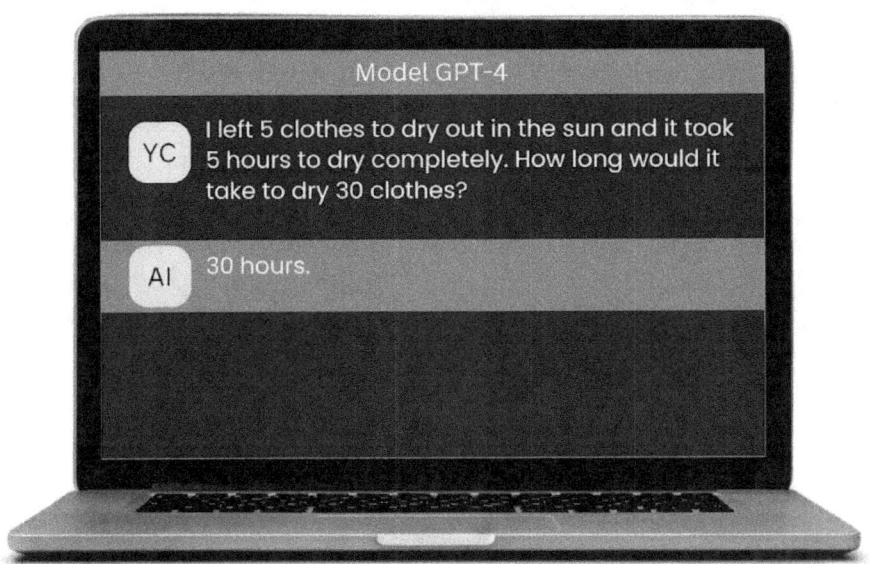

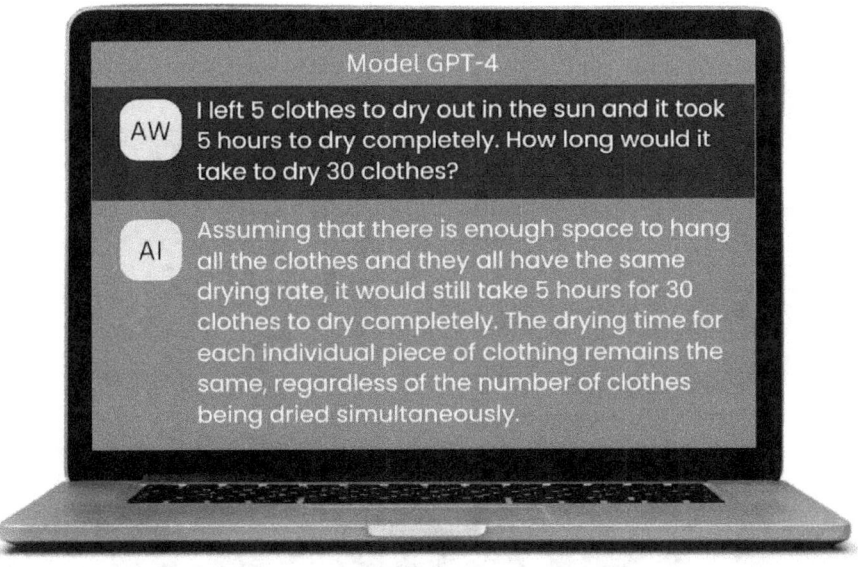

Prompt 2:

(The top image depicts Choi's conversation with OpenAI.
The bottom image shows Alana's same OpenAI prompt with different results.)

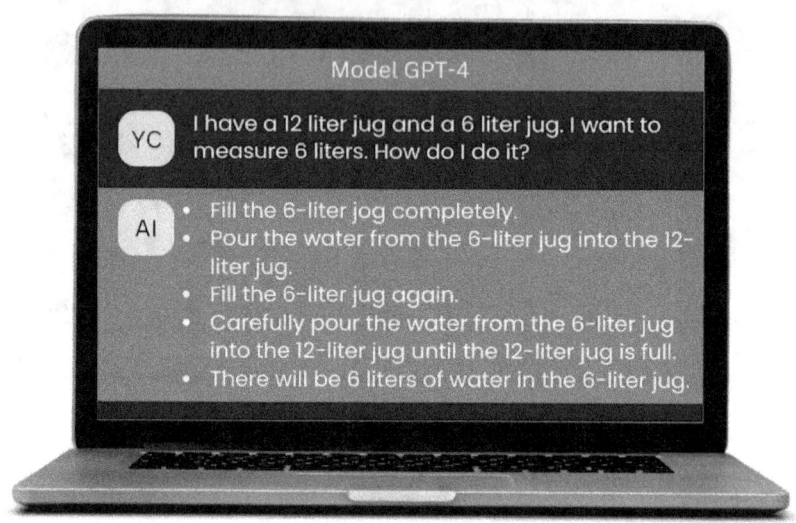

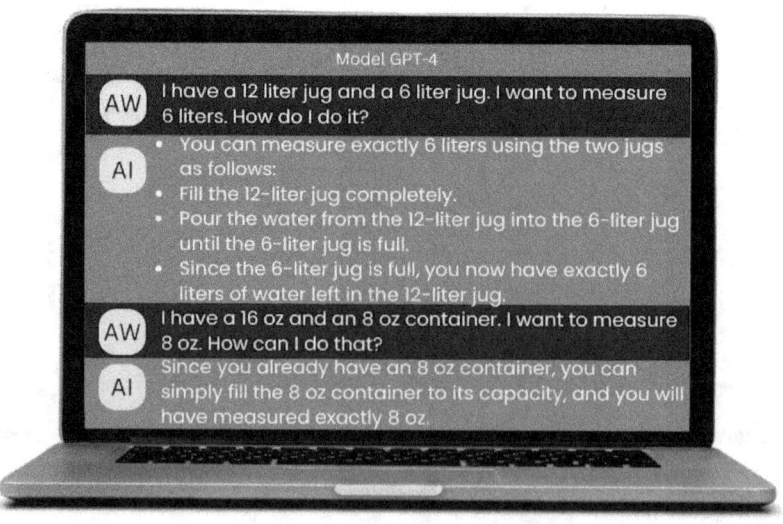

As you can see, when AI doesn't know the answer, it surely acts confident in telling you something that sounds legitimate, which is more commonly being referred to as hallucinations. Based on Yejin's prompts and responses, it is clear that there is currently a significant gap between AI's impressive capabilities and its lack of common sense with regard to understanding basic concepts, context, and common-sense reasoning. However, it is important to note that this will definitely improve over time. Based on my own prompts, I received different responses within the same chat when asking the same questions, highlighting the lack of consistency within the AI-generated content. As we discussed earlier in the book, humans analyze and make sense of the world around us, while AI is (currently) solely predicting and outputting the next most probable word. By raising awareness about these limitations, Yejin encourages us to promote critical thinking in our interactions with AI, and we must instill this within our own students.

Assignment Ideas:
- Provide your students with multiple responses and ask them which was created by AI.
- Ask students to analyze prompt responses to see if they can find incorrect information in the AI-generated content.

Rising Above Ethical Hurdles: Strategies for Success

To navigate the ethical landscape of generative AI in education, educators and administrators can take the following proactive steps; we will cover each of these in greater detail later in the book.

- **Develop comprehensive policies:** Update your school's code of conduct and responsible use policy (covered in chapter 7) to address data privacy, responsible computer-aided assistance, and consequences for academic dishonesty.
- **Educate students and staff to question the output:** Foster a mindset of inquiry and encourage individuals to analyze, verify, and critically assess AI-generated content to cultivate a culture of responsibility.
- **Ensure equitable access:** As we learned during the pandemic, we must strive to provide equal access to all technology, including generative AI tools and resources, for all students, regardless of socioeconomic background or geographic location.
- **Foster a culture of academic integrity:** Educate students, teachers, and parents about the ethical implications of using generative AI in education, emphasizing the importance of original thought and critical thinking. Implement detection tools and strategies, both manual and technology-aided (covered in the next chapter), to identify and address instances of AI-generated content in student work where not appropriate.
- **Promote data privacy and security:** Implement strict data protection measures and ensure third-party AI providers adhere to high data privacy and security standards. Regularly review and update these.

CHAPTER 4

Decoding Dishonesty

Pop Quiz!
Bot or Not?

Directions: Examine each pair; one is composed by a personal correspondent, and the other by artificial intelligence.

1. Should All Drugs Be Decriminalized?
By Harrison Mandell

One of these persuasive essay excerpts was written by 17-year-old Harrison, and the other by OpenAI (2023). Can you guess which is which?

Passage A	Passage B
Drug abuse is a widespread issue that has evolved into a major concern for countries worldwide. The traditional approach of criminalizing drug use has not effectively solved the problem of drug abuse and its impacts...	Drug abuse has become a pressing global concern, and it is clear that traditional approaches of criminalization have fallen short in solving this complex issue. As drug rates continue to rise and communities...

Which pasage was written by Harrison?   A ☐   B ☐

Scan the QR code and complete the form to check your answers.

Pop Quiz!
Bot or Not?

2. Hold My Hand
By Piper Van Schaick

One of these personal narratives excerpts was written by 12-year-old Piper, and the other by OpenAI (2023). Can you guess which is which?

Passage A

I was freaked out. I wanted to be anywhere but here, but no luck. I was just lying there in my mom's lap with snowflakes falling onto my face when these two ski patrollers asked, "Is she ok?"

Passage B

I was in shock. I wanted to be anywhere other than here, but wasn't. I was laying in my mom's lap as snowflakes were falling onto my face when two of the ski patrollers asked, "Is she ok?"

Which pasage was written by Piper?   A ☐   B ☐

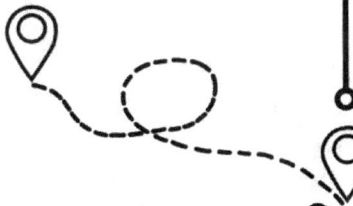

Scan the QR code and complete the form to check your answers.

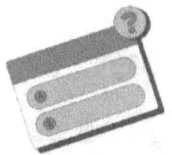

Pop Quiz!
Bot or Not?

3. Can You See Me?
By Zachary Mandell

One of these college admission essay excerpts was written by 18-year-old Zachary, and the other by OpenAI (2023). Can you guess which is which?

Passage A

You might not see me if you were trying to find me in class. I'm a good student, but for a long time, I've been trying my best to stay out of sight - sitting in the back row and hoping the teacher wouldn't say my name..

Passage B

You might not see me if you were looking in my classroom. I am a good student, but for many years, I tried desperately to not be seen - hiding in the back row and praying the teacher wouldn't call my name...

Which pasage was written by Zachary?   A ☐   B ☐

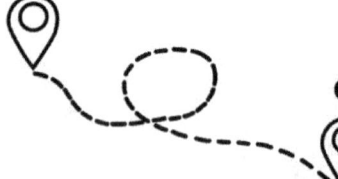

Scan the QR code and complete the form to check your answers.

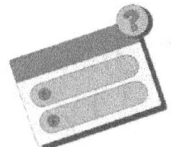

Pop Quiz!
Bot or Not?

4. Technology is Affecting Children's Brains
By Charlie Van Schaick

One of these research paper excerpts was written by 14-year-old Charlie, and the other by OpenAI (2023). Can you guess which is which?

Passage A

There is no way to vanish technology today as it is beneficial to some degree. But children should be children, not robots with their eyes glued to a screen. Start setting goals to look less at your phone and more at the world around you.

Passage B

It is impossible to get rid of technology these days because it helps in so many ways. Kids should be kids and not turn into robots who can't take their eyes off screens. Why not start creating goals to look at your phone less and...

Which pasage was written by Charlie?   A ☐   B ☐

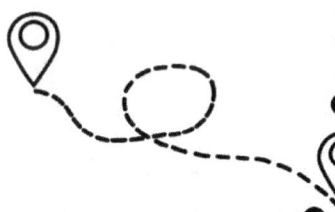

Scan the QR code and complete the form to check your answers.

Pop Quiz!
Bot or Not?

5. Prints
By Hailey W.

One of these masterpieces was created by one-year-old Hailey, and the other by OpenAI (2023). Can you guess which is which?

Exhibit A

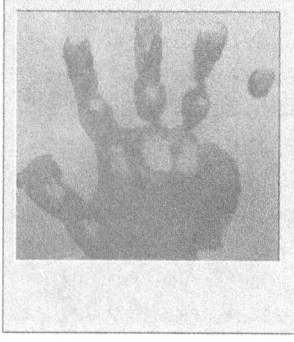

Exhibit B

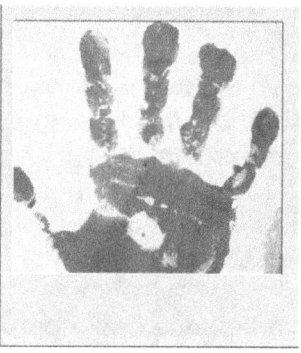

Which masterpiece was created by Hailey? A ☐   B ☐

As you can see, it can be quite difficult to differentiate between the two.

Scan the QR code and complete the form to check your answers.

Shades of Gray

When it comes to academic dishonesty with AI, it isn't black and white; as you can see from the diagram below, there are numerous shades of gray. Miller (2023) showcases a wide spectrum that ranges from entirely human-generated content to completely AI-generated content. The tricky part is deciding where to delineate what's considered dishonest. Determining this boundary, or drawing the line, will likely vary from assignment to assignment, teacher to teacher, and school to school, based on perceptions and ethical considerations. This is why it's crucial to engage in open conversations, set clear expectations, and develop a position. (We will cover all of this later in the book.)

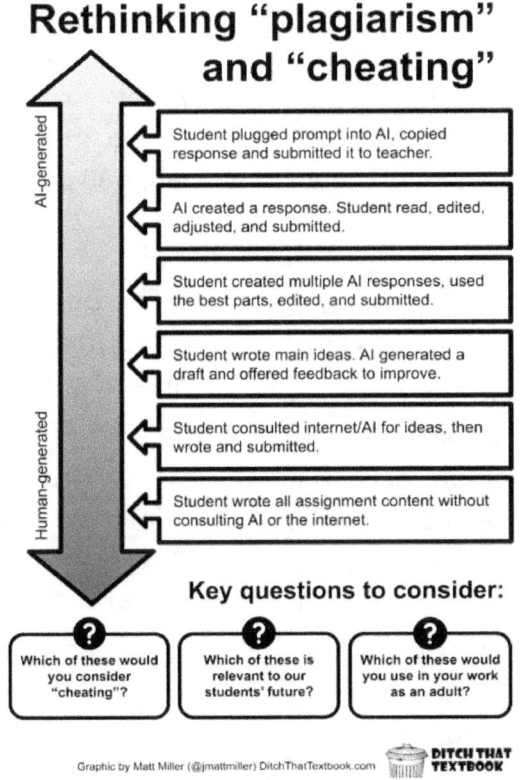

(image used with permission from the publisher)

Shifting Perspectives: Cheating or a Thought Partner?

Let's get cheating out of the way, because the immense benefits and possibilities of AI far outweigh this fear. A neighboring school district's message to their staff was that if you Google plagiarism, you will find "the practice of taking someone else's work or ideas and passing them off as one's own." Who is someone? Someone is a person; AI is not a person. Is it considered plagiarism to use generative AI? Let's ask ChatGPT: When prompted with "Is using generative AI considered plagiarism?" ChatGPT generated the following response: "Plagiarism is the act of presenting someone else's work or ideas as your own without giving proper credit or acknowledgment. This can include copying and pasting text from a source without citation, paraphrasing someone else's work without attribution, or submitting a work that has been created by someone else as your own. Using AI-generated content could be considered plagiarism if you do not disclose that the content was generated by an AI tool and present it as your own original work." (OpenAI, 2023).

Did I use generative AI to help me write this book? You bet I did! You know it's a great thought partner if you've tried it. Generative AI has become my thought partner—I am constantly bouncing ideas off it and demanding revisions ("rewrite this, but add/revise this..."), and 100% of the time, I end up with something much better than I otherwise would have. The purpose of using AI-generated content should be to enhance and support your work, not to replace the need for original thought and effort.

> The purpose of using AI-generated content should be to enhance and support your work, not to replace the need for original thought and effort.

Drawing the Line: Generative AI's Relationship to Academic Dishonesty

We learned that using generative AI isn't technically plagiarism, but what about academic dishonesty? What exactly is academic dishonesty? Academic dishonesty refers to dishonest or unethical actions taken by a student to gain an unfair advantage. By reviewing this definition, we can safely say that the unauthorized use of generative AI falls under academic dishonesty. Using AI-generated content in an academic setting could be considered academic dishonesty if:

1. Assignment guidelines prohibit using AI tools.
2. The student presents the AI-generated content as their own original work without proper citation or disclosing that it was generated by AI.
3. The student uses generative AI to complete assignments, exams, or projects in a manner that violates the guidelines or rules set by the school/teacher.

This will look very different from elementary school all the way through college.

The Originality Conundrum

Because I love to throw you curveballs, let's analyze the ownership of original ideas. What constitutes original thought? Most of our ideas aren't actually our own, they are inspired by things that we've read or heard about. Artists and musicians draw from pieces that they've seen and heard, and they synthesize them into something different. The power of the human brain is its ability to curate and create from our prior experiences.

Companies pay a lot of money for consultants to generate ideas and possibilities. They analyze them and select the ones that align with their mission and vision. What if they can offsource that: ask AI to write 10-20 possibilities and use their human brains to select the ones that most resonate with their goals and ideas. What if you could do that, too? Instead of thinking of your own ideas—or better yet, Googling, researching, synthesizing, and creating your own lesson plan—why not ask AI to write ten ideas for an engaging lesson on climate change, then curate and select the ideas that resonate with you and use AI to help you expand upon them? This is a much more valuable use of your time and energy. As humans, we are amazing curators, and we need to teach our students how to leverage this technology to curate and create as well.

I went back and forth with the idea of authoring this book as "Alana Winnick (in collaboration with Artificial Intelligence)"; it is edgy and timely, but I don't think everyone understands AI just yet. This book reflects my own thoughts and ideas, and I do not want to discredit myself. I interviewed a lot of thought leaders and educators, from the CEO of ISTE (who inspired this section), to an AI Researcher at Microsoft, a professor at the UPenn School of Engineering, the founder of an educational AI software company, multiple school leaders in various roles, and classroom teachers—did I include them as co-authors? No. If I use AI correctly, it is simply a tool to help me curate ideas and make them my own. In fact, one of my own teachers even said, "I read the book! It honestly reads like you're speaking...Ready to use some AI!" This goes to show you that although I used AI to support my work, the end result really is my own voice.

Cracking the AI Code: Detecting Generative AI

As generative AI tools become more sophisticated and accessible, educators and leaders must be prepared to address the potential for misuse and academic dishonesty. There are various techniques, both manual and computer-generated, that can help identify whether a student utilized generative AI.

Manual Evaluation

Educators and leaders should rely on their expertise and intuition to detect possible generative AI usage—trust me, sometimes you'll just know. Signs to look out for include: inconsistencies in writing style, repetition, unusual phrasing or vocabulary, and content that appears unrelated or irrelevant to the topic. Here are two ways to tell if your students are using generative AI:

+ **Revision history:** Check the revision history in Google or Microsoft tools. If sudden large chunks of text appear in a short period of time (copying and pasting), it may indicate AI usage.
+ **Student Monitoring:** Monitor students' work progression over time (remember, it's the process, not the product), both through revisions and in-class assignments—think back to the flipped classroom (we will cover this later in the book): students learn for homework and apply their knowledge by completing assignments in class with teacher supervision and support.

Technology-aided Detection

As we discussed in the beginning of the book, the generated text is simply writing the next most probable word, so if a computer is outputting the next most probable word, then you'd think that it could easily analyze the text to determine patterns and similarities of the words that are strung together. Here, I outline some of the key detection strategies available:

- **AI-based detection tools:** AI detection tools analyze text for subtle patterns and inconsistencies that are characteristics of machine-generated writing and will output a percentage of the likelihood that a generative AI tool was used.
- **Plagiarism detection tools:** As previously discussed, AI-generated content may not always be considered plagiarism, but plagiarism detection software can still be useful for identifying patterns and similarities that may suggest AI-generated content.
- **Stylometry analysis:** Stylometry is the study of linguistic style, which can analyze patterns in writing and identify authorship. There are tools to compare a student's writing style across multiple assignments, helping educators to spot inconsistencies.

Curveball!

In a surprising twist, AI can even outsmart AI! Tools such as GPT-Minus 1 can trick detectors by randomly replacing words with synonyms. Since we know that generative AI predicts the next most probable word, AI can outsmart the system by swapping out words. Imagine a student writes an essay using ChatGPT (achieving a 90% likelihood of AI-use detection score), then applies GPT-Minus 1 to randomly substitute words, resulting in a score of about 10%--astonishing, isn't it? What implications does this have, and how can we guarantee that our students uphold academic integrity? Continue reading to discover more.

Best Practices for AI Detection

According to Ethan Mollick, professor at The Wharton School and author of Co-Intelligence, teachers should be wary of AI detection tools. They have a higher false positive rate than we would like, and I would never want to build a "gotcha" culture on my own district, would you?

Instead, educators and leaders should promote academic integrity and ensure that students are equipped with the skills and knowledge that they need for success in an increasingly AI-driven world. When implementing these tools, educators should consider the following best practices:

- **Ensure fairness and privacy:** When using detection tools, ensure they are being used fairly and consistently across all students, and be mindful of student privacy concerns. Handle cases of suspected AI-generated content with sensitivity, discretion, and due process.

- **Foster a culture of academic integrity:** Detection tools should promote academic integrity rather than solely punishing dishonesty. Encourage open discussions about the ethical implications of AI-generated content and emphasize the value of original thought and effort.

- **Provide guidance and support:** Recognize that students may be tempted to use AI-generated content because they lack understanding, resources, or confidence. Offer guidance and support to help them develop the skills they need to succeed.

- **Set clear expectations:** Communicate your class/school/ assignment guidelines on AI-generated content to students/ teachers, including the potential consequences of violating these policies. Ensure that students are aware of the detection tools being used and their purpose.

- **Use a multi-pronged approach:** Rely on a combination of technology and manual evaluation to increase detection accuracy. Neither method is foolproof, so a multi-pronged approach will provide a more comprehensive assessment.

Pop Quiz! Identifying Academic Dishonesty In Assignments

Directions: For each of the following scenarios involving generative AI in assignments, determine:

A – Do you think the situation would be considered cheating?

B – Would you do this in your own work?

C – Is it relevant to the student's future?

1. A student uses AI to generate a full essay, corrects inaccurate information, and submits the work as their own.

   A – Yes ☐ NO ☐
   B – Yes ☐ NO ☐
   C – Yes ☐ NO ☐

2. A student uses AI to brainstorm ideas for their assignment, develops an outline based on those ideas, and writes the assignment independently.

   A – Yes ☐ NO ☐
   B – Yes ☐ NO ☐
   C – Yes ☐ NO ☐

Scan the QR code and complete the form to see how your answers compare to other readers.

Pop Quiz! Identifying Academic Dishonesty In Assignments

3. A student struggles with a specific paragraph or section of their assignment, uses AI to generate suggestions, and then revises the content in their own words.

A - Yes ☐ NO ☐
B - Yes ☐ NO ☐
C - Yes ☐ NO ☐

4. A student uses AI to paraphrase entire sections of text from various sources and compiles these paraphrased sections into their assignment.

A - Yes ☐ NO ☐
B - Yes ☐ NO ☐
C - Yes ☐ NO ☐

5. A student uses AI to input their original outline and auto generates a presentation with graphics and talking points for an upcoming presentation.

A - Yes ☐ NO ☐
B - Yes ☐ NO ☐
C - Yes ☐ NO ☐

Scan the QR code and complete the form to see how your answers compare to other readers.

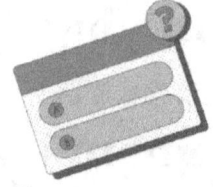

Pop Quiz! Identifying Academic Dishonesty In Assignments

Scan the QR code and complete the form to see how your answers compare to other readers.

6. A student uses dictation to compose passages and then runs them through AI to refine the content: removing filler words, summarizing their ideas, and expanding on different areas.

A - Yes ☐ NO ☐
B - Yes ☐ NO ☐
C - Yes ☐ NO ☐

---

4. A student consults AI to check their work for grammar, punctuation, and clarity, making necessary revisions before submitting the assignment.

A - Yes ☐ NO ☐
B - Yes ☐ NO ☐
C - Yes ☐ NO ☐

---

5. A student uses AI to input their original thoughts and generate a rough draft of their assignment, then reads through the content, heavily revises and restructures the draft, and submits the final version as their own

A - Yes ☐ NO ☐
B - Yes ☐ NO ☐
C - Yes ☐ NO ☐

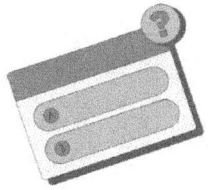

Pop Quiz! Identifying Academic Dishonesty In Assignments

Reflection Questions:

1. Did your responses surprise you?

2. Did the scenarios cause you to rethink your initial views and lead to new insights or perspectives?

3. Responses will vary from person to person. How does that impact your school/class and what can you do to set clear expectations?

Scan the QR code and complete the form to see how your answers compare to other readers.

CHAPTER 5

Navigating Academic Integrity

Giving Credit Where It's Due

It is essential to remember that AI-generated content is not original work and must be cited appropriately. As you've seen throughout this book, I have cited several examples where I used AI-generated text in addition to writing detailed descriptions of how I've used it. This helps to avoid plagiarism, demonstrates integrity and academic honesty, and allows educators and readers to trace the sources of information to verify accuracy.

As long as students properly cite their sources or explain how they used AI, they should be allowed to use it as a thought partner for most writing assignments, depending on what the teacher is looking for. There are many citation guides out there, all of which stress the importance of documenting the exact AI-generated text because, as we learned in "Questioning the AI-generated output," each prompt generates unique responses, even within the same chat.

 # APA Citation Guide

According to the American Psychological Association's (APA) style (2023), the following guidelines are recommended for citing AI-generated text:

Description: In the methods section or introduction of your paper, write HOW you used the tool.

In-text citation: Provide your prompt as well as the relevant AI-generated text:
- **Parenthetical citation:** "Quote" (OpenAI, 2023).
- **Narrative citation:** According to OpenAI (2023)...
- **Example:** When prompted with "How do you properly cite ChatGPT responses in APA format?" I received the response "As of my knowledge cutoff in September 2021, there is no specific guideline for citing AI-generated responses like ChatGPT in the Publication Manual of the American Psychological Association (APA)." (OpenAI, 2023).

 # APA Citation Guide

Reference: This format has been adapted from the software reference template. Example:

OpenAI. (2023). ChatGPT (Mar 14 version)

$\underbrace{\qquad}_{1}$ $\underbrace{\qquad}_{2}$ $\underbrace{\qquad\qquad\qquad}_{3}$

[Large language model].

$\underbrace{\qquad\qquad}_{3 \text{ (continued)}}$

https://chat.openai.com/chat.

$\underbrace{\qquad\qquad\qquad}_{4}$

Reference breakdown:

1. Author = author of model.
    a. Ex: OpenAI
2. Date = Date of the version you used.
    a. Ex: 2023
3. Title = The name of the model.
    a. Ex: ChatGPT (Mar 14 version) [Large language model]
4. Source = URL.
    a. Ex: https://chat.openai.com/chat

 APA Citation Guide

Appendix: You may choose to include the full prompt response in your appendix and refer to it in your paper. Ex: "Quote" (OpenAI, 2023; see Appendix A for the full transcript).

Creative Visuals: As of the date of publication, there are currently no official guidelines for citing AI-generated images or other media.

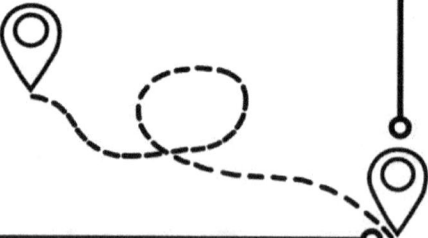

 # MLA Citation Guide

According to the MLA Style Center (2023), the following guidelines are recommended for citing AI-generated text:

Quotation example: When asked how to properly cite ChatGPT responses in MLA format, ChatGPT noted, "As of my knowledge cutoff in September 2021, there is no specific guideline for citing AI-generated responses like ChatGPT in the Modern Language Association (MLA) Handbook." ("Proper MLA citation").

Paraphrase example: ChatGPT's current knowledge cutoff is September 2021, so it does not have specific guidelines for citing AI-generated responses ("Proper MLA citation").

 # MLA Citation Guide

Work cited–list entry:

"Proper MLA citation for ChatGPT responses"
1

prompt. ChatGPT, 14 Mar. version,
1 (cont.)   2   3

OpenAI, 14 May 2023,
4   5

chat.openai.com/chat.
6

Reference breakdown:

- Author = Treating the tool as an author is not recommended.
1. Source = Describe what was generated. Ex: "Proper MLA citation for ChatGPT responses" prompt.
2. Container = The AI tool. Ex: ChatGPT
3. Version = Specific version. Ex: 14 Mar. version
4. Publisher = The name of the company. Ex: OpenAI
5. Date = The date the content was generated. Ex: 14 May 2023
6. Location = General URL of the tool. Ex: chat.openai.com/chat

 # MLA Citation Guide

Creative Visual Works:

Fig. 1. "Visual of artificial intelligence"
1

prompt, DALL-E, version 2,
1 (cont.)   2   3

OpenAI, 14 May 2023, labs.openai.com/.
4   5   6

Reference breakdown:
 1. Create a description of the prompt
 2. Name of AI tool
 3. Version
 4. Publisher
 5. Date created
 6. Location

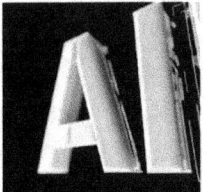

Fig. 1.

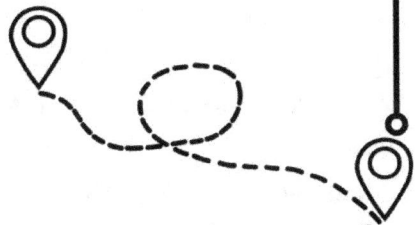

AI Etiquette 101: Upholding Academic Integrity

If you are using generative AI and want to avoid plagiarism, it is important to:

- **Attribute the source:** Mention that the content was generated by an AI tool and provide proper citation if necessary.
- **Combine content with original thoughts:** Use the AI-generated content as a starting point or reference, but incorporate your ideas, analysis, and understanding to create a well-rounded and original piece of work.
- **Review and edit the content:** AI-generated content may not always be accurate or appropriate. Review and edit the generated content to ensure it meets your requirements and accurately reflects your understanding or perspective on the topic.

As Ms. Sinclair continued to explore the possibilities of generative AI in her classroom, she found herself engaging in positive conversations in the faculty lounge about her experience. Her colleagues were excited to see her join the conversation and impressed with some of the ways that she had used it. One day, Ms. Sinclair noticed that her colleague, Hesitant Mr. Huberman, seemed reluctant to embrace AI. He was worried that his students would cheat on their assignments by copying and pasting content straight from the chat. Ms. Sinclair acknowledged his concerns. "I understand why you might feel that way. I had the same concerns at first," she confessed. "Over the past few weeks, I've discovered when used responsibly and ethically, generative AI can actually serve as a thought partner for both us and our students."

Ms. Sinclair went on to explain how AI could help students brainstorm ideas, refine their writing, and even overcome writer's block. "If students are encouraged to use AI as a sounding board for their ideas and cite the tool appropriately, it can actually enhance their learning experience. You can ask them to attach their chat history along with their final draft to make you feel more comfortable that they used the tool appropriately," she said. Mr. Huberman raised an eyebrow, intrigued, but he still seemed somewhat hesitant. Ms. Sinclair offered a suggestion. "Why don't you give it a try. You might be surprised by the results." As they finished their lunch and headed back to their classrooms, she hoped she planted a seed of curiosity within her colleague. Would he be willing to give AI a chance, or would his hesitation hold him back? Only time will tell.

CHAPTER 6

Privacy Playbook
Safeguarding Student Data

Careful Consideration

As we learned previously in the book, generative AI tools offer tremendous potential for assisting with IEPs and other highly personalized learning materials; however, we must carefully consider the data we put into these systems. What happens to the data, how it is deleted, and who sees it are all questions we must consider; the need to maintain data privacy and security becomes paramount. While educators' intention may be to assist their students, the exposure of student data can result in significant damage. This chapter explores the challenges and best practices for leveraging generative AI in a manner that respects students' privacy, while still delivering individualized educational plans.

Balancing Personalization and Privacy

When using generative AI and other data-driven tools to create personalized student plans, educators must strike a balance between harnessing the power of data and protecting the privacy and security of students' personal information. The following best practices can help educators achieve this balance:

- **Proprietary/intellectual property:** Information that you provide in your prompts may be used by the company for their own purposes, such as training the model, without consent. As I always tell teachers and students, assume nothing is private—if you put it out there, it's out there.
- **Redact personally identifiable information (PII):** Remove all PII, such as names, addresses, or Social Security numbers to prevent unintended disclosure of sensitive information.
- **State and federal laws and regulations:** Adherence to laws and regulations is essential, especially if AI was used to make data-informed decisions and recommendations.
- **Train and educate students and staff:** Provide regular training on the importance of data privacy and security, as well as best practices for handling data responsibly.

Responsible Personalization: Creating Student Plans While Respecting Privacy

With data privacy and security measures in place, educators can confidently use generative AI to create personalized student plans following these strategies to maintain student privacy:

- **Focus on aggregate data:** Use aggregated and anonymized data to help you identify patterns and trends, without revealing individual students' identities.
- **Leverage AI responsibly:** Ensure that the tools you use are designed with privacy in mind.

- **Monitor and adjust:** Regularly review and evaluate the effectiveness of AI-generated learning plans and make adjustments as needed, remember, you know your students best.
- **Utilize non-PII data:** Rely on non-PII data, such as the students' learning style, interests, or performance of certain tasks.

One day, Ms. Sinclair overheard a conversation in the faculty lounge. Mr. Lou C. Datta had been using AI to generate IEPs for his students. Ms. Sinclair was excited to see her colleague embracing AI, but when she learned that Mr. Datta had been including his students' full names in his prompts, she was concerned.

Knowing that this violated their state's data privacy and security regulations, Ms. Sinclair approached him and explained the potential danger of including PII in generative AI tools. She explained to him that, "Including a child's personally identifiable information in a tool that does not comply with our state's regulations could result in a data breach, or even worse, identity theft. It's important to prioritize the privacy and security of our students."

Mr. Datta did not realize the potential consequences of his actions. He replied, "Thanks for bringing this to my attention, I didn't realize the potential harm and I'll make sure to be more cautious in the future." Ms. Sinclair was relieved that her message got through to him. Fostering a culture of caution and responsibility around the use of technology was very important to her.

CHAPTER 7

Socrates' Standards
Developing a Position Statement and Policies that Reflect Your Mission and Values

Setting Expectations

It's inevitable that students and teachers will utilize generative AI in their work—and I really hope they do. But what's considered acceptable? To maintain a culture of academic integrity in The Generative Age, it is essential to have clear and precise policies in place. These policies should outline the class/school's stance on AI usage in student work, provide guidelines for acceptable use, and consequences for violating the policy. This way, your school/class is prepared to address the challenges and opportunities posed by generative AI, fostering a culture of academic integrity and responsible technology usage for students and educators.

Crafting a Position Statement

A position statement for using generative AI in education contributes to a thoughtful and proactive approach to integrating AI into education while upholding school/class values and ethical commitments. Here are several reasons why schools need to develop a position statement for the use of generative AI:

- **Consistency and coherence:** Ensures that all stakeholders, including teachers, students, and parents, understand the expectations and guidelines for using AI in education.
- **Guiding principles:** Establishes guiding principles and values that shape the school's approach to using generative AI to help create a shared understanding of your vision and goals.
- **Responsible use:** Outlines the ethical considerations and responsible practices related to generative AI use in education and includes guidelines on data privacy, academic integrity, and accessibility.
- **Stakeholder engagement:** Encourages open dialogue and collaboration among various stakeholders, which promotes a sense of shared ownership and responsibility for the ethical use of generative AI.

Developing a position statement on using generative AI in education requires collaboration with your administrative team, and input from stakeholders including teachers, administrators, students, parents, and the wider community. Put students in the driver's seat to let them think about questions related to AI ethics and responsible use of technology while helping build awareness about the technology they already use. Include their input into your statement. Consider the following steps as you work together to create a comprehensive and effective statement:

1. Schedule a series of meetings or workshops to discuss the implications of generative AI in education, including its potential benefits and risks.
2. Encourage open and honest dialogue among stakeholders, emphasizing the ethical considerations surrounding AI usage.
3. Develop a shared understanding of the role of generative AI in education, considering the balance between technology usage and the importance of fostering original thought and creativity.
4. Draft a position statement that reflects the school's stance on generative AI usage, ensuring it aligns with the school's mission, values, and educational goals.

An example of a school's position statement can be found within the Chappaqua case study in chapter 9.

Homework: Analyzing the Effectiveness of Existing Policies

Directions:

- Take a look at your policies. The first step in addressing generative AI in student work is to review your existing class policies, code of conduct, and acceptable use policy.
- Crafting a policy exclusively around technology isn't recommended; policies should be viewed as guiding and overarching principles.
- Consider the following questions as you assess your current policies:

1. Count the number of Don'ts in your policy. Count the number of Do's. What do you notice? It's important to reframe your policies to be positive.

 # Homework: Analyzing the Effectiveness of Existing Policies

2. Do your policies address academic honesty? (Notice how I didn't say academic dishonesty, plagiarism, or cheating?).

3. Do your policies include authorized computer-based collaboration? Keyword: authorized. Include examples of what's acceptable. Remember, reframe it to the positive.

4. Are your policies general enough in their technology and digital resources coverage, including AI tools? You need not refer to any specific technologies.

A+ Policies: Revising to Address the Use of Generative AI

To ensure that your policies effectively address the use of generative AI in student work, consider implementing the following revisions:

- Reframe your policy to be more positive.
- Ensure that academic honesty is included in your policy.
- Clearly define technology's potential applications in the context of academic work.
- Specify acceptable uses of technology in student work.
- Establish a process for reporting and investigating the suspected use of generative AI in student work and what consequences you will consider.

Examples of policy additions:
TeachAI.org looks like a promising resource. A group of highly coveted collaborators, including ISTE, ASCD, AASA, code.org, OpenAI, and Microsoft, just to name a few, are currently working on policy recommendations, guidelines, and best practices for using AI in education; sign up to stay up-to-date as things develop.

CHAPTER 8

∼

Rebooting Education
2.0 Strategies for Teaching in The Generative Age

Passport to Success: Equipping Students for The Generative Age

With the emergence of generative AI, we face both new opportunities and new challenges that require us to adapt our approach to teaching. We need to rethink what's important to teach our students. By incorporating these essential elements into our teaching, we can better equip students with the skills and mindset that they need to navigate and thrive in the future.

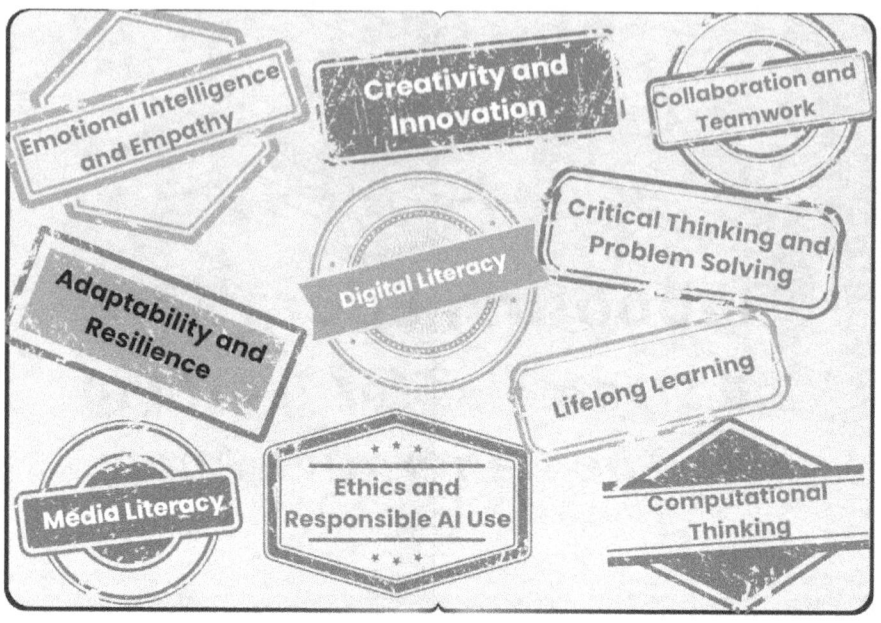

Process over Product: Creating More Meaningful Assignments

If a bot could be used to easily complete activities/assessments, you need to rethink how you assess learning; remember, it's all about the process, not the product. When Google became mainstream, I said "If your question is 'Google-able,' it's not a good question." With gener-ative AI, the same idea applies: if the assignment is "AI-able," it's not a good assignment. My 17-year-old cousin said, "ChatGPT has single-handedly destroyed education" because his assignments are completely "AI-able." It doesn't need to be this way. We need to rethink the types of assignments we are asking of our students and revisit the tried-and-true educational frameworks that have been proven to work, but adding a new AI-twist.

Generative AI-Proofing Assignments

As we progress into The Generative Age, we need to remind ourselves that if the assignment is "AI-able," it's not a good assignment. Not sure if your assignment is AI proof? Run it through an AI tool yourself before asking students to complete it. We should be asking students to apply their knowledge and show us that they comprehend the content. Asking for rote facts is no longer important. Remember back in the days when we were quizzed on rote memorized facts such as dates of specific events? In today's world, with a smartphone in our pocket and knowledge at our fingertips, we no longer need to memo-rize facts. We need to think about what's important in The Generative Age and re-design our lessons and create authentic assessments to help students thrive in this new era— which can be uncomfortable for many educators. Our tasks should provide real world applications. How are you going to do that? If you're an administrator, what type of profes-sional development are you going to offer? We will cover professional development in the next chapter.

Original Thought and Creativity

As discussed earlier in the book, it's important to balance AI assistance and human intervention. We must emphasize the value of original thought and creativity and encour-age teachers and students to balance between utilizing technology to enhance their work and preserv-ing their perspectives and voices. Promoting responsible and critical usage of AI tools can help students

> Promoting responsible and critical usage of AI tools can help students develop valuable skills that will benefit them academically and in their future careers.

develop valuable skills that will benefit them academically and in their future careers.

Analyzing Student Prompts

Math teachers often don't always care about their students getting the correct answer; instead, they often care more about the student showing their work and HOW they got to the answer. The same goes with generative AI—we should care more about the student's thinking process. With generative AI, if you change how you ask the question, you get a different response; how you phrase the prompt is important. The more perceptive the prompt, the more discerning the response; I will provide a prompt engineering framework later in this chapter. Looking at the chat history will give you a sense of how the student thinks and will allow you to learn more about the student.

Some AI tools allow users to export or create shareable links to their prompts, which can then be submitted to the teacher for review, along with the completed assignment, similar to the research note cards that some teachers ask students to turn in with their papers. This will allow the educator to see the original prompts and how the student modified the content to reflect their own voice. Pretty soon I suspect there will be an educational platform that ties the chat history right into the assignment for teacher review.

Assignment idea: Going back to the importance of the human element and AI balance, students can be asked to reflect on how AI assisted them in developing their assignments, and how they included their own original thoughts in the process. Writing a reflective piece about how certain aspects of the output influenced their work can help students avoid perceiving AI as a tool for cheating.

The SAMR Model

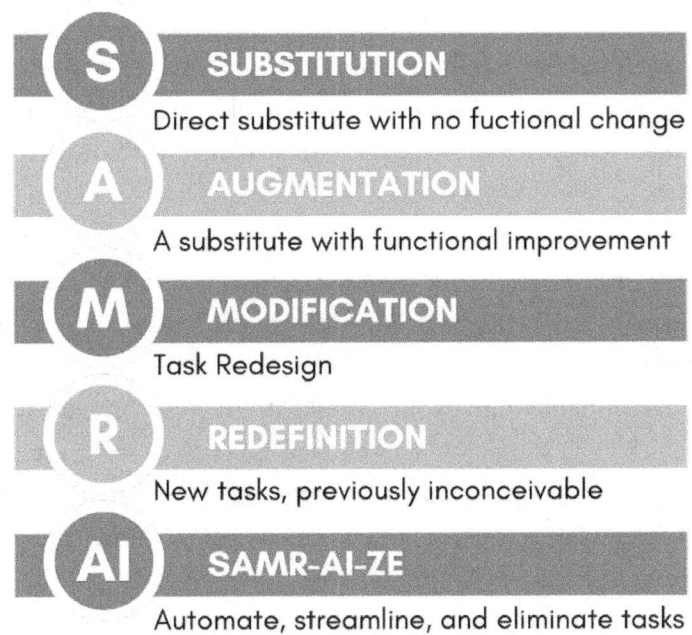

S SUBSTITUTION
Direct substitute with no fuctional change

A AUGMENTATION
A substitute with functional improvement

M MODIFICATION
Task Redesign

R REDEFINITION
New tasks, previously inconceivable

AI SAMR-AI-ZE
Automate, streamline, and eliminate tasks

According to The University of Arizona Center for Assessment, Teaching, and Technology (2021), "SAMR, is an educational technology framework developed by Ruben R. Puentedura, Ph.D. in 2010, to help educators think about effective technology integration. SAMR is an acronym that describes a range of functions of technology in education: Substitution, Augmentation, Modification, Redefinition." (para 1). "These first two parts of the framework are seen as enhancements, not changing the basic function of the information, but improving it with the application of technology...The SAMR model refers to Modification and Redefinition as a Transformational process, meaning it actually alters the learning process and potentially the learning outcomes, as a result." (para 4-5). Here's how generative AI can be utilized at each level of the SAMR model to enhance the educational experiences:

- **Substitution:** A direct substitute for traditional tasks without significant functional improvements. Example: Create a simple quiz for students, replacing traditional quiz-making methods - but remember, we should be rethinking this type of assessment!
- **Augmentation:** Augment existing tasks by providing functional improvement. Example: provide students with real-time feedback on their essays or an AI-tutor to offer personalized support.
- **Modification:** Modify tasks, making them more efficient and effective. Example: Using AI to generate content into multiple formats (text, visuals, audio) to accommodate diverse learning styles.
- **Redefinition:** Create new tasks that were previously inconceivable. Example: Cross-curricular project that integrates a skills-based curriculum and leverages an AI-tutor to provide customized resources and insights to facilitate the learning process.

- **New category alert: SAMR-AI-ZE!!** How about unofficially creating a new category, which I am coining SAMR-AI-ZE. Like the precision and skill of a samurai sword, AI allows us to automate, streamline, and eliminate, or slice through, tasks that are redundant, time-consuming, or do not significantly contribute to the learning process. I first heard about this unofficial category from AJ Juliani, who emphasizes the transformative potential of AI in redefining the role of education.

Bloom's Taxonomy

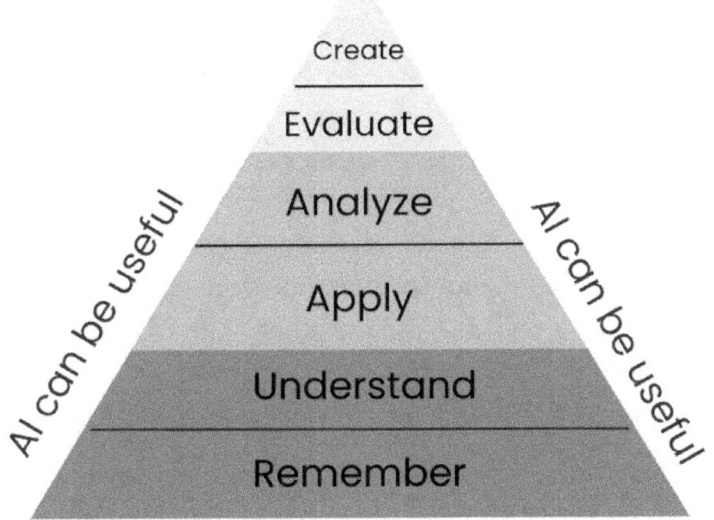

According to Vanderbilt University's Center for Teaching (2010), Benjamin Bloom, along with other collaborators, published a framework called Taxonomy of Educational Objectives, which categorizes educational goals into six major categories: Knowledge, Comprehension, Application, Analysis, Synthesis, and Evaluation. This framework became more popularly known as Bloom's Taxonomy and has been applied by educators at all levels, from Kindergarten to higher education (para 2).

Each level builds upon the previous one, starting with basic recall of information and building up to the ability to generate new ideas and create new knowledge. At the lower levels of Bloom's Taxonomy, generative AI can be extremely useful. For example, you can use AI to generate multiple choice questions or summarize text; however, when it comes to the higher levels of Bloom's Taxonomy, generative AI falls short. While AI can create new content, it cannot create abstract thinking, ideas, or mental models that don't exist yet (keyword: yet). At this moment, it also lacks the ability to truly understand context and make judgments based on subjective criteria—but that will very likely change as the technology evolves.

This has important implications for education. If we are pushing students to higher levels of Bloom's Taxonomy—specifically, their ability to create new knowledge—then, depending on the purpose of the assignment, perhaps they should no longer be required to do the lower-level tasks that generative AI can quickly and easily perform for them. This frees up time for students to engage in more complex, creative thinking. It is still, and always will be, important for students to develop their own thinking skills, and to be able to create original ideas that do not yet exist. This requires a focus on abstract thinking, creativity, and mental modeling. As educators, it is our responsibility to push students to these higher levels of thinking and to provide them with the tools they need to develop their own original ideas.

Universal Design for Learning

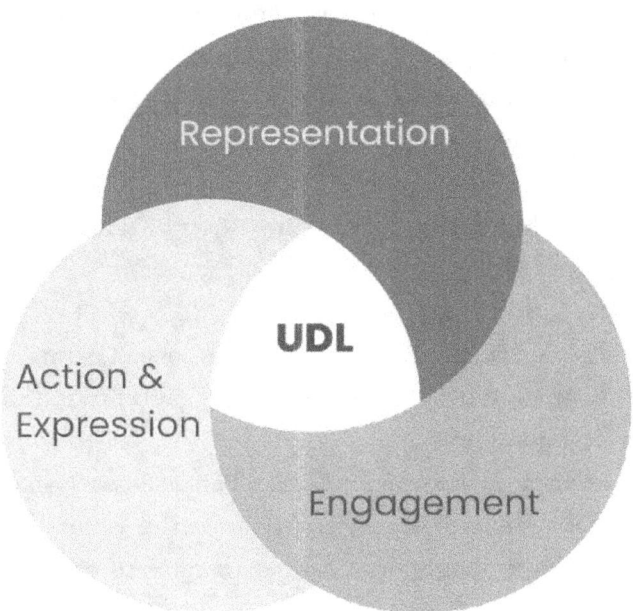

According to Cornell University's Center for Teaching Innovation "Universal design for learning (UDL) is a teaching approach that works to accommodate the needs and abilities of all learners and eliminates unnecessary hurdles in the learning process. This means developing a flexible learning environment in which information is presented in multiple ways, students engage in learning in a variety of ways, and students are provided options when demonstrating their learning." (para 1). The idea is that diversity among learners is the norm rather than the exception and encourages educators to design educational experiences to accommodate diversity from the very start. UDL is built on three main principles that guide the design of learning experiences. Here is how generative AI can support each:

- **Representation:**
  - **Generate content in multiple formats:** Automatically convert text into audio or visual formats to accommodate different learning needs and preferences.
  - **Simplify complex text:** Rewrite complex texts into simpler language, making it easier to understand for learners with diverse reading abilities.
  - **Translate content:** Translate educational materials into multiple languages, ensuring that language barriers do not hinder learning.
  - **Create visual aids:** Generate diagrams, illustrations, or infographics to help visualize complex concepts and ideas.
- **Action and expression:**
  - **Adaptive assessments:** Create personalized assessments that adapt to each learner's performance. Offer questions of varying difficulty and format to cater to different learners' strengths.
  - **Alternative assessments:** Generate a variety of assessment types, such as written assignments, multimedia presentations, projects, or interactive quizzes. This provides voice and choice and allows learners to demonstrate their understanding in ways that suit their preferences and strengths.
  - **Automated feedback:** Provide instant, personalized feedback on learners' work, helping them identify areas for improvement and guiding them through the learning process.
- **Engagement:**
  - **Personalized learning pathways:** Create tailored learning experiences that cater to individual needs and interests.
  - **Collaborative learning:** Facilitate collaboration among learners by generating prompts, questions, or scenarios that encourage discussion and problem-solving.

◇ **Motivational content:** Generate content that is relevant, engaging, and enjoyable for learners.

The Flipped Classroom

The Flipped Classroom

teachers are more actively involved in **the learning process.**

The likelihood of reliance on AI decreases when...

The flipped classroom approach has the potential to address concerns surrounding the use of generative AI. According to Harvard University's Derek Bok Center for Teaching and Learning " A flipped classroom is structured around the idea that lecture or direct instruction is not the best use of class time. Instead, students encounter information before class, freeing class time for activities that involve higher order thinking." (para 1). By implementing this model, educators can devote more time closely observing students in class as they work on their assignments and watch their work develop. The flipped classroom model involves the following key components:

1. **Pre-class prep:** Students review instructional materials, such as videos, readings, or lecture notes, before attending class. Perhaps these materials are created using generative AI?

2. **In-class activities and assignments:** With foundational knowledge acquired before class, students can focus on applying and deepening their understanding during class time.

3. **Teacher observation and support:** As students work on assignments, teachers can closely monitor their progress. This allows educators to see the work develop in real-time and to provide immediate feedback.

4. **Personalized attention:** This model enables teachers to spend more one-on-one time with students as they work on assignments.

5. **Minimizing AI reliance:** With teachers more actively involved in the learning process, the likelihood of students relying on generative AI for completing assignments decreases. The flipped classroom approach encourages students to develop their own critical thinking and problem-solving skills, rather than depending on AI-generated content.

Project-Based Learning

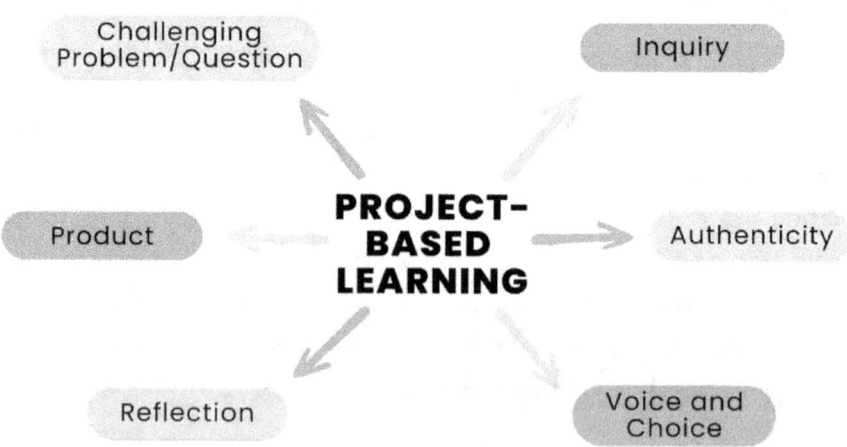

In February 2023, Education Week published an article titled *Outsmart ChatGPT: 8 Tips for Creating Assignments It Can't Do*. At the time, educators were highly concerned about students cheating, so this article aimed to prevent that. Tip number seven suggested implementing project-based learning because it makes it more difficult for students to cheat, or solely utilize AI-generated content to complete assignments. "ChatGPT was never going to do that project for them, South said. "It'd be impossible to cheat on that project with it. And the kids didn't want to cheat because they were doing something really cool and interesting and relevant to their lives." (para 24) (Klein, 2023)

According to Boston University's Center for Teaching and Learning, "Project-based learning (PBL) involves students designing, developing, and constructing hands-on solutions to a problem. The educational value of PBL is that it aims to build students' creative capacity to work through difficult or ill-structured problems, commonly in small teams." (para 1). In PBL, students explore and investigate complex issues, such as the UN sustainable development goals, which emphasizes the development of essential skills such as critical thinking, problem-solving, creativity, communication, and collaboration.

I had the pleasure of interviewing AJ Juliani, a leading project-based learning expert, on *The Generative Age* podcast. I will summarize the information here, but I invite you to listen to the entire episode.

Scan here to watch the AJ Juliani Generative Age episode:

AJ discusses the need for a shift from standardized testing towards performance tasks and project-based learning. As AI continues to advance, PBL will become an increasingly valuable tool for students to demonstrate their learning and understanding. By leveraging generative AI within the context of PBL, educators can create engaging and relevant learning experiences that capture students' interests, push the boundaries of traditional classroom activities, and prepare students for the future workforce.

By incorporating generative AI tools into project-based learning, educators can create more engaging, personalized, and effective learning experiences for their students while promoting ethical and responsible use of technology. This will allow us to shift the focus away from cheating and towards constructive and productive relationships with technology. By encouraging students to explore AI as a creative partner in PBL, educators can promote responsible and ethical use of AI in the classroom.

Generative AI can complement PBL in several ways:

- **Develop essential skills:** Develop crucial skills such as teamwork, digital literacy, data analysis, and critical thinking, which are essential for success in today's world.
- **Develop critical thinking and evaluation:** Assess the reliability, accuracy, and biases of AI-generated content.
- **Engage diverse learners:** Create inclusive learning environments by providing resources and support for students with different learning styles, abilities, and backgrounds.
- **Enhance creativity:** Students can use AI as a starting point for their projects, inspiring new ideas and perspectives and building upon them to create original work.

Design Thinking

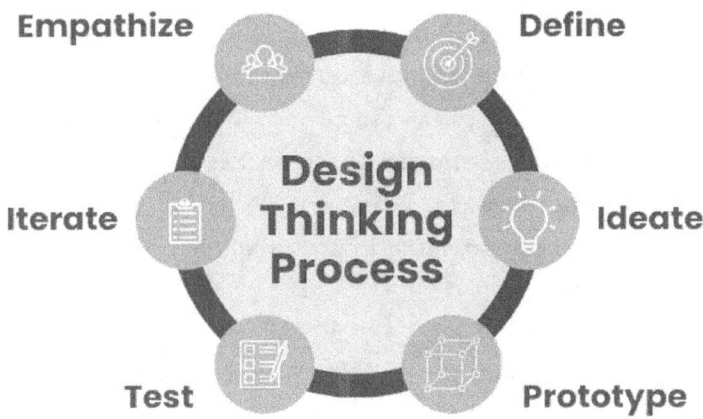

According to MIT's Sloan School, "Design thinking is an innovative problem-solving process rooted in a set of skills." Linke, R. (2017, para 1). By integrating this concept (outlined below) into education, we can create engaging learning experiences that foster critical thinking, creativity, collaboration, and problem-solving skills, which will empower students to become active participants in their own learning journeys and equip them with the mindset and skills needed to thrive in The Generative Age. Here's a breakdown of how generative AI can support each level of the design thinking process:

- **Empathize:** Help students understand and empathize with their target audience by generating insights into user behavior, preferences, and challenges.
- **Define:** Assist with finding patterns and trends in the data, which will help students articulate the problem statement. Generative AI can also help them write out the problem statement.
- **Ideate:** AI can serve as one of the best idea generators, producing a vast and wide range of design concepts and creative possibilities, encouraging students to think outside of the box. An

individual student may come up with just two ideas, but generative AI can come up with 20+ ideas in seconds.

+ **Prototype:** Create mock-ups and models based on their chosen design ideas, which will save students time and help them focus on refining their ideas and incorporating user feedback.
+ **Test:** Provide real-time feedback on the prototype, identifying areas for improvement.
+ **Iterate:** Identify potential improvements and generate alternative solutions.

It's important for your students to know that even though they are young, they can make a big difference. A recent example of this includes students in my own school district breaking way too many of our new Microsoft Surface devices. After interviewing students and believing that they were not mistreating their devices, I complained to Microsoft numerous times. They tested the devices and put them through the design thinking process (using our data) and determined that if they placed just a few more glue drops behind the touchscreen, it would significantly decrease the breakage rate. We then discussed why I wouldn't purchase the model with the kickstand that didn't have this issue. I said that it isn't sturdy enough for students to put on their lap while deeply working in contorted angles all over the classroom, so they asked me if we wanted to work with a case company to help design a different case that would meet the needs of young students. How cool is that? Students in my district will learn that not only did they have an impact on the design of their current device, they will have an opportunity to use the design thinking process to create a case that will actually go into production. I suspect AI can't do that from start to finish! Your projects need not be this big; they can be as small as creating an eco-friendly bird feeder or 3D designing a door opener.

Teaching Your Students About AI and Prompt Engineering

At this point, I hope I have convinced you of the importance of AI in our future. Similar to how we teach students about cybersecurity, it is paramount to educate them about AI so that they can critically examine its role in their own lives and make informed decisions about its use. By teaching generative AI and prompting in schools, you will promote technological literacy, help students explore creativity and innovation, encourage critical thinking and computational thinking skills, introduce ethical considerations, and–most importantly–prepare students for future job opportunities that do not yet exist.

Teaching students about proper prompting is essential in The Generative Age because it helps them understand how to formulate clear, concise, and thoughtful prompts which will help them achieve more accurate and relevant responses. By understanding how to frame ques-tions and prompts effectively, students will develop critical thinking and problem-solving skills, further developing their analytical abilities.

Not sure how to properly prompt? Daniel Fitzpatrick, Amanda Fox, and Brad Weinstein, authors of *The AI Classroom* came up with the fol-lowing prompt engineering framework to help you and your students get started. (Fitzpatrick et al., 2023)

PROMPT ENGINEER FRAMEWORK

@AmandaFoxSTEM
@DanFitzTweets
@WeinsteinEdu

First, PREP the Machine.

1

P Prompt
Introduce the question with a prompt

2

R Role
Give it a role or voice

3

E Explicit
Be explicit in your instructions

4

P Parameters
Set the parameters of the answer

Then, EDIT the Output.

5

E Evaluate
Evaluate your AI output content for language, facts, and structure

6

D Determine
Determine accuracy and corroborate with source.

7

I Identify
Identify biases and misinformation in output.

8

T Transform
Transform content to reflect adjustments and new findings

Re-PREP & EDIT until satisfied

The #AICLASSROOM

(image used with permission from the publisher)

102

Scan here for tutorials:

Assignment idea: Ask students to try different variations of a prompt and then write a reflection on what made it a good prompt to get the best response.

Curricular Resources: Below are just a few curricular resources and pre-made lessons to help you get started with engaging your students in the exciting world of AI:

aiEDU: A non-profit that creates equitable learning experiences that build foundational AI literacy.

- ✦ Cost: Free!
- ✦ aiedu.org

Code.org: AI isn't magic... it's code! Learning how AI is changing the ways we live, work, and learn.

- ✦ Cost: FREE!
- ✦ Code.org/ai

Day of AI: Empower your students with the knowledge and skills to flourish in a world with AI.

- ✦ Cost: Free!
- ✦ Dayofai.org

ISTE Educator Guides: Hands-On AI Projects for the Classroom Guides provide curricular resources about AI across various grade levels and subject areas, including "unplugged" activities.

- ✦ Cost: FREE!
- ✦ iste.org/areas-of-focus/AI-in-education

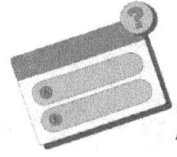

Pop Quiz! Is This Assignment AI-Proof?

Directions: For each of the following scenarios, determine:
   A – Is this assignment AI-proof?
   B – Are the skills relevant to the student's future?

1. A teacher moderates a Socratic seminar, fostering an open and respectful environment that encourages critical thinking and deep analysis of a literary work. Students are required to analyze the text, prepare thoughtful questions, engage in respectful dialogue, and provide textual evidence to support their arguments.

   A - Yes ☐ NO ☐   B - Yes ☐ NO ☐

2. Students write a persuasive essay arguing for or against the implementation of renewable energy sources in your community.

   A - Yes ☐ NO ☐   B - Yes ☐ NO ☐

Scan the QR code and complete the form to see how your answers compare to other readers.

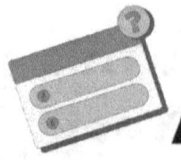

Pop Quiz! Is This Assignment AI-Proof?

3. During independent work time, students engage in one-on-one discussions with the teacher. Each student is individually prompted, and their responses include examples of material covered over the semester.

   A - Yes ☐ NO ☐   B - Yes ☐ NO ☐

4. Students work in groups to create a multimedia presentation on one of the United Nations' sustainable development goals, including research, data analysis, and a proposed solution. They are assessed on their collaborative skills and individual contributions.

   A - Yes ☐ NO ☐   B - Yes ☐ NO ☐

5. Students design a hands-on science experiment about the unit content they are learning in class. They conduct the experiment, record their observations, and present their findings.

   A - Yes ☐ NO ☐   B - Yes ☐ NO ☐

Scan the QR code and complete the form to see how your answers compare to other readers.

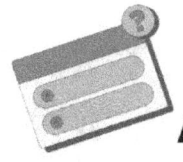

Pop Quiz! Is This Assignment AI-Proof?

Scan the QR code and complete the form to see how your answers compare to other readers.

6. Students produce and edit a podcast episode centered around an interview they conducted. To complete this assignment, they conducted research on their interviewee, crafted insightful questions, facilitated the interview, and edited the audio.

   A - Yes ☐ NO ☐   B - Yes ☐ NO ☐

7. Students create a digital storyboard illustrating a scenario related to a sensitive topic, such as bullying, digital citizenship, or mental health, followed by leading a classroom discussion on their thoughts, experiences, and emotions, creating a safe space for open dialogue.

   A - Yes ☐ NO ☐   B - Yes ☐ NO ☐

8. Students write and deliver a TedTalk for the larger school community. Their talks are based on their own personal experiences, capture their audience's emotions, and incorporate reflections for the future.

   A - Yes ☐ NO ☐   B - Yes ☐ NO ☐

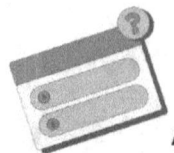

Pop Quiz! Is This Assignment AI-Proof?

9. Students write an essay on technology's role in transforming teaching and learning. They must include multiple viewpoints and cite at least three sources.

   A - Yes ☐ NO ☐   B - Yes ☐ NO ☐

10. Students complete a questionnaire on their academics, career, and personal goals. They reflect on their interests, strengths, and challenges, research potential career paths, and create a digital portfolio showcasing their skills and achievements. Educators offer personalized advice and guidance based on the students' self-assessments and portfolios, fostering individual growth and development.

    A - Yes ☐ NO ☐   B - Yes ☐ NO ☐

Scan the QR code and complete the form to see how your answers compare to other readers.

By examining the "AI-proof-ability" and relevance of assignments, teachers can create engaging and meaningful learning experiences. These experiences will emphasize real-world connections and the development of valuable skills that will help students thrive in an increasingly AI-driven world. Remember to keep questioning, adapting, and evolving your approach to teaching as the landscape continues to change.

Remember to keep questioning, adapting, and evolving your approach to teaching as the landscape continues to change.

Homework: Asking Difficult Questions and Having Difficult Conversations

Reflection Questions:

1. How are we currently educating students?

2. What do students need to be learning?

3. Rethink the content. Can we perhaps utilize more of an interdisciplinary approach?

4. How are we asking students to showcase their knowledge? How should this shift in the coming years?

5. What is the future of assessments in a world influenced by AI?

Homework: Asking Difficult Questions and Having Difficult Conversations

6. What skills are important for students to know in order to be successful in the present?

7. Our students will grow up and work in The Generative Age. What skills do they need to be successful in the future?

8. How will the role of a teacher transform over the next decade?

9. What will define an educated person in ten years from now?

10. How can we effectively prepare students for the future we don't yet understand?

CHAPTER 9

From Chalkboards
to Chatbots

Professional Learning for Disruptive Technology

Supporting Educators in The Generative Age

In preparation for this book, I met with my own superintendent, seeking to understand his perspective on generative AI. Surprisingly, his concern was not about his own understanding of the technology, but more about ensuring how we can best support our staff. Generative AI is not going away and if we don't teach educators how to effectively leverage these tools, it will be a disservice to our students.

When used appropriately, in the first phase, generative AI can free up valuable time, enabling educators and leaders to focus on fostering relationships that drive the class/school's success and allows teachers to concentrate on their students' learning. In the second phase, described in chapter 1, students will truly be immersed in their own highly personalized and engaging learning plans with their teachers as mentors and facilitators. Early adopters will undoubtedly embrace this technology;

however, the resistors usually need more convincing. One way for you to get these teachers on board is demonstrating the time-saving benefits. Bring them together for a hands-on exploration, and you'll likely witness their astonishment as they discover the possibilities.

Giga-bits of Knowledge: Effective Professional Learning Techniques

To effectively incorporate generative AI (or any disruptive technology) into the curriculum, one must provide educators with targeted and ongoing professional development; there cannot be a disconnect between how teachers are trained and how they are expected to work. This will ensure that they have the

> There cannot be a disconnect between how teachers are trained and how they are expected to work.

knowledge and skills needed to integrate AI effectively into their teaching practices, and ultimately enhance the learning outcomes for their students.

Just like students in a classroom, there is no one-size-fits-all approach to implementing professional development in your schools, and you will see four very different case studies below to prove it. It is dependent on leadership styles and culture. The beauty of generative AI is that as long as there is access, it levels the playing field—no matter the budget, this technology is free and widely accessible. Here are some strategies for achieving this:

- **Addressing concerns and fears:** Providing reassurance, delivering comprehensive information, and promoting an open dialogue will help to demystify AI and foster a more comfortable and confident integration.

+ **Encourage collaboration and sharing:** Providing opportunities for educators to collaborate and share best practices can help create a continuous learning and improvement culture. Witnessing their colleagues' success provides evidence, a sense of inspiration, and encouragement, which reaffirms the belief that, "If they can do it, so can I."
+ **Hands-on training:** Educators should have opportunities to practice using AI tools in real-world settings to develop their skills and gain confidence in incorporating them into their teaching practices. This experiential approach allows teachers to see the benefits firsthand.
+ **Ongoing training:** As mentioned earlier, AI is a rapidly evolving field, and educators need ongoing training to stay up-to-date with the latest tools and techniques. This can be achieved through faculty meetings, workshops, conference days, and online training.
+ **Ongoing emails/newsletters:** Include resources and updates to keep the conversation going.
+ **Teacher agency:** Offer diverse opportunities for learning, including voice and choice in how teachers access knowledge.

Byte-sized Learning: Hands-on AI Exploration for All Educators

By providing comprehensive professional development that addresses generative AI's practical applications and ethical considerations, educators can better understand the technology's potential and integrate it into their everyday work. While early adopters are eager to explore the possibilities, some educators may be more resistant to change. This section offers strategies for providing professional development to all educators.

- **Establish a common language and baseline understanding:** Begin professional development sessions with an overview of generative AI, its capabilities, and its potential impact on education. Encourage thoughtful conversations and address any concerns or misconceptions.

- **Highlight immediate benefits and real-life applications:** Help educators visualize the potential benefits by providing examples of real-life applications (refer back to "A Universe of (AI) Possibilities" in chapter 1 for examples) and ensure that they leave with at least one concrete, immediately practical use. Demonstrate how generative AI can save them time and effort. By experiencing this firsthand, even resistant educators may become more open to integrating generative AI into their work.

- **Offer opportunities for exploration and play:** Provide interactive demonstrations and opportunities to play and explore. This will help educators identify potential applications in their work and encourage adoption among resistant educators.

- **Involve early adopters in professional development planning and facilitation:** To create a more robust and engaging professional development agenda, involve early adopters in the planning process. These educators can share their experiences and insights, offer peer-to-peer support, and help generate enthusiasm among their colleagues.

- **Encourage critical thinking and skill development:** Promote a critical lens by discussing the ethical implications of generative AI and its potential impact on students. Help educators identify the skills they want to teach their students and explore how generative AI can serve as a tool to support skill development.

Continue reading to see four vastly different case studies that showcase successful generative AI implementation strategies that you can apply to any disruptive technology. Depending on your leadership style and culture, pick and choose what resonates with you and your district's philosophy—and if your district is doing something different that's working, please reach out and let me know; I am always eager to learn from and with you!

Case Study: Pocantico Pride

Once upon a time, there was an Educational Technology Director in a small suburban school district who was super passionate about generative AI...yes, that would be me! After the release of ChatGPT, I stormed into my superintendent's office explaining what I discovered. I knew that we had to get in front of it. We initially decided to present at a faculty meeting about the opportunities and implications of generative AI for teacher use (as there was not a version that was compliant with New York State student data privacy and security laws), followed by hands-on exploration. Unfortunately, when we pulled out our calendars, we quickly realized that with all of our existing initiatives, we didn't have available faculty meetings to dedicate to this new initiative.

As a result, when I first launched The Generative Age podcast, it was simply a webinar; I sent out an email to staff members, letting them know that I would be presenting an informational and voluntary session through The New York State Association for Computers and Technologies in Education (NYSCATE), and they were more than welcome to attend. I was pleasantly surprised by how many staff members joined at 7:00 pm on a Monday night! The following day I sent out an email with the recording so those who were interested could watch. Staff members who attended were enthusiastic and went back to school the next day, raving to their colleagues about what they had learned.

It began with small, seemingly inconsequential conversations with teachers and staff. "Did you write that French reading comprehension passage? I heard about this AI tool that could help..." or "Have you tried using AI to help you come up with lunch specials using the ingredients you have on hand?" These subtle mentions of the technology and job-embedded PD disrupted the status quo, slowly weaving generative AI into the district. I made it a point to purposely bring it up and showcase its capabilities every time I saw someone create something that could have been aided by AI.

Similar to a train, most new initiatives require a considerable amount of energy to set them into motion. Overcoming challenges such as resources, buy-in, and rollout plans are necessary to lay the groundwork, or tracks, for success. Comparable to how it takes minimal effort for a train to switch tracks, new initiatives can be more easily steered once they're in motion. This has been my outlook on most new initiatives until I spoke to a fellow district leader who told me about the law of entropy, which was definitely not in my vocabulary. He explained that in simple terms, it takes a lot less energy to create disorder than it does to create order. He went on to say that our goal is actually to disrupt the system, and disrupting the system is relatively easy. Creating order, which will come later on, is the part that requires more energy. Think about this: how much time does it take you to mess up your bed vs making your bed? A lot less time, right?

This disruption—the creation of disorder—was achieved with minimal energy. Much like messing up a bed is easier than making it, introducing generative AI this way was far more simple than organizing a formal launch event at a faculty meeting. Teachers and staff were encouraged to explore the tool, make a mess (keeping student data secure, which is always a top priority in our district), and push the tool to the edges of its capabilities. We followed up with a faculty meeting at the end of the year, during which teachers were encouraged to share their findings with their colleagues. This playful, safe environment facilitated an intersection around best practices, without the pressure to follow any specific set of rules.

We made a conscious effort to keep this organic, lighthearted, and enjoyable, knowing that the uphill journey towards full integration of generative AI too early could lead to resistance. Instead, we focused on creating a culture that felt comfortable with experimentation and exploration. Allowing teachers and staff to determine its strengths and weaknesses on their own terms encourages a sense of ownership and investment in the technology. This approach fostered a community that was eager to share their discoveries and learn from one another's experiences.

And though there will undoubtedly be challenges in the future, the foundation has been laid for a meaningful integration into our district. All it took was the simple, subtle power of conversation and the law of entropy to spark a change that would impact our district for years to come. Additional professional learning opportunities include:

- **Ongoing informational emails**: Information and resources, including my NYSCATE Generative Age podcasts. Adding "in collaboration with AI" to any emails where it is used in order to promote it as an acceptable thought partner.
- **Faculty meeting**: Initial faculty meeting involved teachers sharing their findings and experiences. A more formal faculty meeting with stations facilitated by ISTE Certified educators, similar to Pearl River's approach, which you will read in the next session, followed.
- **Breakfast club**: Once a month, teachers could opt to come in for a district-funded breakfast before school to "talk shop" and share their latest discoveries.
- **In-service professional development**: Will be provided over the summer.
- **Conference day**: Teachers will lead sessions on AI curricular integration during the opening conference day.

- **Community education:** We will host an evening event to educate parents and community members about artificial intelligence and its opportunities and implications on education.
- **Ongoing job-embedded PD:** Meaning, I will continue to be annoying!

Case Study: Pearls of AI Wisdom

Pearl River School District quickly realized the complex nature of generative AI and how intimidating it can be for educators who are unsure of how to harness its power in their classrooms. Recognizing this challenge, a team of two instructional tech coaches and one literacy coach, led by Director of Technology, Jamie Haug, embarked on a mission to demystify AI and present it in a positive light for their teaching staff.

Adopting a low-stakes, accessible approach, the team organized a full-day fair in a dedicated room that resembled station rotation, inspired by Catlin Tucker, but with a more flexible format. Staff members were encouraged to drop in at their convenience for as little as five minutes, with no sign-up required. This informal setting closely resembled vendor booths at a conference and put teachers at ease, allowing them to engage with the AI tools and resources on display without feeling overwhelmed.

The event featured various stations, each with a unique AI-related theme, designed to have a non-overwhelming impact on attendees. The team cleverly utilized ChatGPT to generate creative names for each station, adding a touch of fun to the experience. To further encourage open discussion and address any concerns, a "parking lot" was set up for teachers to submit questions and share their thoughts.

Stations

The room was set up using a circular layout, allowing participants to explore the stations in any order they wished. Each station featured a specific AI tool or application, including:

+ **AI Teaching Assistant:** This station demonstrated how AI can save teachers time and reduce workload by automating tasks like writing college reference letters. Participants learned how to use templates and prompts to generate personalized content with ChatGPT.

+ **Overlord's Guide to Good Grades and Great Ethics:** Addressed concerns about academic integrity and responsible use of AI tools in education.

+ **Parking Lot:** In addition to the main stations, a Google Form parking lot allowed staff members to submit questions.

+ **Perfect Prompts:** Focused on teaching staff how to effectively prompt ChatGPT for precise and targeted responses, ensuring they get the most out of the AI tool.

+ **Virtual Sidekick for the Classroom:** Showcased ways to create instructional content using ChatGPT for more engaging and dynamic learning experiences.

+ **Wakelet Resource Hub:** provided further reading and material on AI in education.

Staff members who attended the event found value in the time-saving and instructional applications showcased, and some expressed interest in using the tools immediately following the event. A follow-up survey or discussion board was suggested by participants as a way to keep the conversation going and share ideas among colleagues.

The most popular piece of feedback that Pearl River received from attendees was that AI was much different than they thought it would be. The team overheard comments such as "I am going back upstairs to try it out" and "You just saved me a few hours." Overall, Jamie and her team felt that the day was a huge success. She stated, "If you could see the look on some of our staff members' faces when tinkering with some of the prompts, you would agree."

Lessons Learned and Future Improvements

While the event was successful, I asked Pearl River for suggestions for future iterations, and they recommended having "Ask me about AI" stickers to promote conversation and further engagement as well as conducting a follow-up session to share experiences and collect feedback.

Overall, the Pearl River School District's generative AI professional development initiative demonstrates an easy and non-threatening way to showcase the benefits of integrating AI tools into the educational setting. By providing practical, hands-on experiences and facilitating open discussion, the district has set the stage for ongoing innovation and improvement in teaching and learning.

Scan here for additional information and materials:

Case Study: Croton's Community Connection

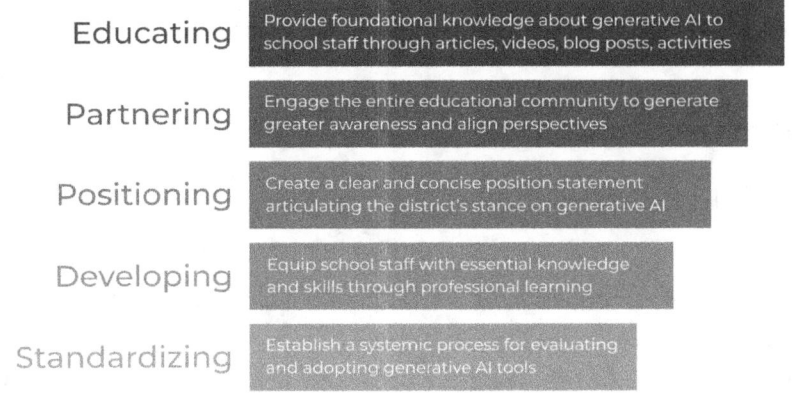

(used with permission from Dr. Cameron Fadjo)

During a visit from Dr. Bill Daggett back in March 2022, the Croton-Harmon Union Free School District was forewarned that generative AI had the potential to disrupt the educational system. Dr. Daggett emphasized the need to prepare for the upcoming changes and specifically predicted generative AI's relevance in the educational setting. As predicted, generative AI started appearing in the news just a few months later. Croton had more time than most to prepare for this disruption, and very carefully strategized their implementation. Initially, there were concerns and fears associated with generative AI, as people tend to fear the unknown; however, their goal was to build awareness and turn that fear into opportunity.

Board of Education Buy-In

Croton held a public work session with their board of education trustees in a modified "Think Tank" protocol, which was recorded and made available to the community. Through this process, community members gained a better understanding of AI's potential impact on the education system. Their main focus was to build awareness of generative

AI, in addition to its potential impact on teaching practices and student learning. Before they could move forward, it was crucial to have the support and understanding of their community.

Community Buy-In

Croton has a strong commitment to fostering a supportive and informed community. It's important for their parents and community members to comprehend the reasons behind incorporating new technologies, like AI, into their teaching practices. By seeking their understanding and involvement, they aim to create a partnership that aligns professional learning with community support. How do you think they launched their community involvement? They invited Dr. Daggett back for another presentation, this time to their community.

Innovation Team

Croton quickly formed an Innovation Team, composed of students, faculty, parents, administrators, and community members. This team continued the conversation and developed recommendations for the district. They discussed how generative AI could impact the classroom, shaping their teaching practices and assessments, while exploring the best ways to integrate AI into their school system. An example of this includes questioning the relevance of the traditional five-paragraph essay when students can rely on generative AI to compose it, which helped them recognize the need to adapt their practices to accommodate this new technology.

Professional Development Plans

At the time of publication, Croton had not conducted any official professional development specifically related to generative AI. Knowing that the implementation of AI will have significant impacts on their assessment practices and the way that they integrate generative AI into everyday tasks, they believed that diving into professional development immediately wasn't the right timing, and that's OK. Each school system has different levels of readiness and capacity.

However, they have big plans for the summer, which will include various topics, ranging from innovative pedagogical practices, to the role of generative AI in the classroom. They aim to differentiate their training to meet teachers where they are, aligning with Chappaqua's philosophy, which you will read about later on.

Croton Harmon Disruptors

Croton is currently working on forming a team called the CHUFSD Disruptors, which will operate based on Disruptive Innovation Theory. This theory explains the fears and unknowns associated with the use of new technology as well as the steps associated with normalizing the technology. This team will guide Croton's implementation of all technology that may initially feel uncomfortable. The objective is to overcome that discomfort through understanding, hands-on practice, and ultimately integrating disruptive technology into their system. The influence of this team will spread throughout their educational system by turn-keying their experiences and takeaways to the larger community.

As you can see, by educating parents and staff and addressing their fears and concerns, Croton was able to remove some of the apprehensions and build a supportive environment which will flourish in the years to follow.

Case Study: Chappaqua Champions Change

In January 2023, just two months after the release of ChatGPT, Chappaqua Central School District recognized the rapid growth and potential impact of generative AI. They boldly embraced the future of education by implementing a professional learning program centered around this technology, light years before most schools publicly addressed the topic. I enjoyed discussing the district's groundbreaking rollout on the The Generative Age podcast; I will recap it here, but I invite you to listen to the entire episode.

Scan here to watch the Chappaqua Generative Age episode:

The district developed guiding principles, position statements, and rapidly developed two distinct learning paths for educators: exposure and awareness (superintendent's conference day), and deeper learning (in-service courses). Each path was tailored to meet the diverse needs of the district's teaching staff, ensuring that educators of all experience levels could engage.

In-depth learning path: Optional in-service courses to help early adopters dive deeper into understanding generative AI and its potential applications.

Exposure and awareness path: A more general professional development program to help all educators become familiar with the technology and its implications.

The In-Depth Learning Path

The district implemented an in-service course in January 2023, which included four sessions that focused on understanding the technology, exploring its applications in the classroom, and developing lesson plans that integrate generative AI. This path aims to provide educators with a comprehensive understanding of the technology and its potential applications in the classroom, allowing them to integrate AI into their teaching practices effectively. It consists of a series of sessions that cover various aspects of generative AI. Some key components of this path include:

- **Developing lesson plans with AI integration:** The third session focuses on creating lesson plans that incorporate generative AI. Examples include using AI-generated text for podcasts or generating ideas for essays.
- **Evaluation and reflection:** In the final session, educators evaluate their learning experience and share insights about the integration of generative AI in their classrooms.
- **Mapping learning and classroom integration:** In the second session, educators explore how they can use generative AI in their classes, discussing its potential applications across different subject areas.
- **Understanding generative AI:** Educators learn about the technology behind generative AI, how it functions, and why it is important. This session was held in January and focused on its effects on education, and the need for its integration.

Exposure and Awareness Learning Path (Superintendent's Conference Day)

In March 2023, Chappaqua hosted a full-day Superintendent's Conference Day on the topic of artificial intelligence and one of the key elements that made their professional learning program an example of best practices was learner "voice and choice." The day began with a self-assessment, allowing teachers to identify themselves as apprentices, practitioners, or maestros in generative AI. This self-assessment enabled the district to offer differentiated sessions tailored to each educator's level of experience with AI and allowed for a tailored approach that catered to each individual's expertise and interests.

1. **Apprentice**: Beginner AI user looking for some foundational knowledge and skills
2. **Practitioner**: For those who have tried it and have begun to integrate AI into their work-life.
3. **Maestro**: Experienced AI user. Ready for a high-level conversation on AI application in schools.

This learning path focuses on providing a general overview of the technology, enabling educators to become familiar with AI and its potential impact on their teaching practices. This day aimed to provide educators with an engaging and informative experience, helping them understand and embrace generative AI as a valuable tool in their teaching practices. Key sessions and activities include:

+ **AI revolution video presentation:** A video by John Spencer titled "The AI Revolution" was shown to provide further context on the growing importance of AI in education and society at large.
+ **Breakout choice sessions:** Choice sessions followed, including titles such as "Cheating in the Digital Age: Navigating the Intersection of Artificial Intelligence and Academic Integrity," "More than ChatGPT: Explore, Tinker & Learn about other AI tools" and "Explore How to Use ChatGPT in your Classroom: Amping Up Your Classes & Lesson Planning."

- **Demonstration of GPT:** Educators were given the opportunity to log into ChatGPT with their devices and explore the platform, allowing them to gain firsthand experience with the technology.
- **Disruptive technology in education:** A mini-lesson was conducted to discuss the concept of disruptive technology in education and how AI compares to previous disruptive innovations such as calculators and spell-check.
- **Expert panel discussion:** Featuring experts from various fields, including banking, medicine, and education. These experts shared their experiences with AI development and its applications in their respective industries. They also discussed the implications of AI in education and its potential impact on future professions, emphasizing the importance of AI literacy for students and educators alike.
- **Gallery of AI-generated works:** Gallery of items created using ChatGPT. Participants were asked to guess whether a human or AI generated each item. This activity aimed to showcase the impressive abilities of generative AI and stimulate discussions about its potential applications in education.
- **Student led professional learning:** Chappaqua's #TruthSquad, a team of trained high school students, also prepared and ran student-led teacher discussions on topics such as "The Future of AI Grading," "AI: Does it Increase or Inhibit Learning?" and "Reprogramming our Learning."

The Superintendent's Conference Day concluded with a reflection on the insights gained throughout the event and the district's vision for generative AI in education. The day's activities provided a solid foundation for educators to understand the technology and its potential applications in their classrooms.

Chappaqua's Position Statement (Working Draft)

As a language model designed to assist with a variety of tasks, ChatGPT(AI/Machine Learning) can be a valuable tool for middle school students in their academic pursuits. However, it is important to use ChatGPT appropriately and ethically to ensure that it does not become a means of academic dishonesty.

ChatGPT should be used as a resource to support learning and exploration. It can be used to clarify concepts, generate ideas, and provide additional information on a topic. Middle school students can also use ChatGPT to practice writing and language skills by engaging in conversations with the model.

However, it is important to note that ChatGPT should not be used as a replacement for learning to write well, for critical thinking or for independent research. Simply copying and pasting answers generated by ChatGPT without understanding the underlying concepts or doing any additional research would be academically dishonest and could result in negative consequences.

In addition, ChatGPT should not be used to cheat on assignments or assessments. Using ChatGPT to generate answers for exams or homework assignments without proper attribution would be a violation of academic integrity and could result in disciplinary action.

Therefore, it is recommended that middle school students use ChatGPT as a tool for learning and exploration, but also recognize its limitations and use it ethically. Teachers should also provide clear guidelines for appropriate use of ChatGPT and educate their students on academic integrity and the consequences of plagiarism.

Chappaqua's Guiding Principles of Success

+ **Embracing AI as an educational tool:** The district fosters a mindset of embracing AI as a valuable educational tool, rather than banning or discouraging its use. This approach allows students to make mistakes and learn about AI's responsible and ethical use under the guidance of their educators.

+ **Encouraging continuous learning:** The district is committed to providing ongoing professional development opportunities for educators to stay informed about advancements in AI and its applications in education, ensuring that teachers are well-equipped to incorporate the technology into their teaching practices.

+ **Ensuring AI literacy:** The district emphasizes the importance of AI literacy for both students and educators, recognizing that a comprehensive understanding of AI's capabilities and limitations would be crucial for success in various professional fields.

+ **Policies for AI integration:** A critical aspect of their initiative was the development of a district-wide position statement. These guiding principles outline the district's stance on embracing, teaching, and enhancing education through generative AI, ensuring that all stakeholders are aligned in their understanding and approach to the technology.

Next Steps and Future Initiatives

With the position statement and professional development experiences, the district will continue its AI professional learning initiative through focused in-service courses and professional learning sessions. These efforts will address generative AI applications in specific content areas, allowing educators to explore tailored strategies for incorporating AI into their subject matter.

Chappaqua Central School District's proactive approach to integrating generative AI into their professional learning program demonstrates their commitment to keeping up with evolving technologies in the field of education. By engaging staff in hands-on learning experiences and promoting a deeper understanding of AI's potential impact, the district is well-positioned to adapt and thrive in the rapidly changing educational landscape.

Scan here for additional information and materials:

CHAPTER 10

Commencement of Change
Taking the Reins of Education's Future

As we reach the end of our journey together (for now) into the world of generative AI in education, it's time to turn the page from theory to practice. The future of education lies in your competent hands, and the decisions you make today will shape the learning experiences of tomorrow. Your leadership and vision guides your school/class toward a future that responsibly and effectively embraces AI. You hold the power to transform the educational experience and unlock the potential that generative AI holds for shaping a brighter future for both education and the generations to come.

> The future of education lies in your competent hands, and the decisions you make today will shape the learning experiences of tomorrow.

Take a moment to reflect on the insights, strategies, and best practices you have gained throughout this book. Consider how you can implement these ideas, tailoring them to meet the unique needs of the educators, students, and community you serve. This is not just a time for contemplation, but a call to action. The future of education calls for visionary leaders like you who are prepared to navigate the uncharted waters of AI integration. Your determination and foresight will shape how we teach, learn, and grow in the ever-evolving Generative Age.

As you rise to the challenge, remember that you are not alone. Connect with fellow educators, leaders, and educational technology experts to share ideas, experiences, and support. Foster a collaborative environment that embraces change and empowers everyone involved to participate actively. I invite you to stay current and continue the conversation by tuning into The Generative Age podcast, and following me on social media (@AlanaWinnick) for updates. Additional information can be found at www.AlanaWinnick. com and www.GenerativeAge.com

The future of education is now in your hands. Foster an environment of learning and curiosity, and empower your students to thrive in the AI-driven world of tomorrow. Seize this moment, and, together, let's embrace the possibilities that await us in The Generative Age.

References

Armstrong, P. (2010). Bloom's Taxonomy. Vanderbilt University Center for Teaching. Retrieved [May 30, 2023] from https://cft.vanderbilt.edu/guides-sub-pages/blooms-taxonomy/.

Bloom, B. (1984). The 2 Sigma Problem: The Search for Methods of Group Instruction as Effective as One-to-one Tutoring. *Educational Researcher*, *13*(6), 4–16.

Boston University Center for Teaching and Learning. (n.d.). *Project-Based Learning: Teaching Guide*. Project-Based Learning: Teaching Guide | Center for Teaching & Learning. https://www.bu.edu/ctl/guides/project-based-learning/

Choi, Y. (2023). *Why AI is Incredibly Smart and Shockingly Stupid*. TED. https://www.ted.com/talks/yejin_choi_why_ai_is_incredibly_smart_and_shockingly_stupid/c?language=en

Fitzpatrick, D., Fox, A., & Weinstein, B. (2023). *The AI Classroom: The ultimate guide to artificial intelligence in education*. TeacherGoals Publishing.

Harvard University Derek Bok Center for Teaching and Learning. (n.d.). *Flipped classrooms*. https://bokcenter.harvard.edu/flipped-classrooms

How do I cite Generative AI in MLA style?. MLA Style Center. (2023, April 12). https://style.mla.org/citing-generative-ai/

Khan, S. (2023). *How AI Could Save (Not Destroy) Education*. TED. https://www.ted.com/talks/sal_khan_how_ai_could_save_not_destroy_education/c

Klein, A. (2023, February 15). *Outsmart CHATGPT: 8 tips for creating assignments it can't do*. Education Week. https://www.edweek.org/technology/outsmart-chatgpt-8-tips-for-creating-assignments-it-cant-do/2023/02

Linke, R. (2017, September 14). *Design thinking, explained*. MIT Sloan. https://mitsloan.mit.edu/ideas-made-to-matter/design-thinking-explained

McAdoo, T. (2023, April 7). *How to cite chatgpt.* American Psychological Association. https://apastyle.apa.org/blog/how-to-cite-chatgpt

Miller, M. (2023). *AI for Educators Learning Strategies, teacher efficiencies, and a vision for an artificial intelligence future.* Dave Burgess Consulting.

OpenAI. (2023). ChatGPT (Mar 14 version) [Large language model]. https://chat.openai.com/chat

Samr and quintet: Essential Tools for Effective Technology Integration. The University of Arizona Center for Assessment, Teaching, and Technology. (2021, November 24). https://ucatt.arizona.edu/news/samr-and-quintet-essential-tools-effective-technology-integration

Smith, M. (2022, November 22). *"It Killed My Spirit": How 3 teachers are navigating the burnout crisis in education.* CNBC. https://www.cnbc.com/2022/11/22/teachers-are-in-the-midst-of-a-burnout-crisis-it-became-intolerable.html

Taylor, Stephen (2023). *(If You) USEME-AI Model.* The Western Academy of Beijing https://sjtylr.net/if-you-useme-ai/

U.S. Department of Education, Office of Educational Technology, Artificial Intelligence and Future of Teaching and Learning: Insights and Recommendations, Washington, DC, 2023.

Universal Design for Learning: Center for Teaching Innovation. Universal Design for Learning | Cornell University's Center for Teaching Innovation. (n.d.). https://teaching.cornell.edu/teaching-resources/designing-your-course/universal-design-learning

Acknowledgements

My family and friends: My parents, Andrew, Keri, Jackie, Carrie, aunts, uncles, cousins, Patricia and beyond! Thank you for all of the ongoing love, support, and encouragement you've provided me with throughout my life. For giving me space when I need it, while being incredibly supportive at the same time. I know that you'll always be there to ride the waves—celebrate the highs and hold my hand for the lows. Seriously, thanks for always putting up with me. I love you!

- **The Generative Age community:** I am immensely grateful to NYSCATE, Cameron, The Generative Age guests, live participants, and listeners who have made both the podcast and creation of this book possible. Your engagement, feedback, and diverse perspectives have shaped the narrative and infused this book with invaluable insights. Your enthusiasm and support have been a constant source of motivation. Together, we will "keep learning and keep growing in...The Generative Age." See you soon!

- **The NYSCATE community:** Amy, Mary Beth, Sean, the board, and all of our members- your passion and enthusiasm is what keeps me going! From my early days as a member, to becoming a presenter/facilitator, and eventually a board member, this journey has been nothing short of extraordinary. I am eternally grateful for the opportunities and relationships formed throughout the years and none of this would have been possible without you. Thank you!

- **Pocantico:** Rich (Superintendent of the year), Mike, Adam, Christy, Gina, Deb, staff, students, board of education, and community. Thank you for taking a leap of faith and entrusting me with a leadership role. Your belief in my abilities and the autonomy that you provide me with has been instrumental in my journey. Thank you for supporting my crazy ideas and fostering an environment that encourages innovation and growth.

- To all our dedicated teachers, I am truly grateful for your patience, flexibility, and willingness to step outside of your comfort zones when I know all too well that the challenges of our profession can be very overwhelming. I hope that this book and its concepts will inspire you to reclaim your time and continue making a difference in the lives of our students. As always, if you want to try something new, please know that I am always available and eager to collaborate with and support you.

- **Reviewers:** A huge thank you for your time and effort! I am deeply grateful for the invaluable insights, constructive criticism, and unwavering support that you have provided me with. Your honest feedback has been instrumental in refining this book to ensure that it resonates with my readers in addition to helping me grow and evolve as a writer. You know who you are, and I want you to know that I chose each of you for a reason: I deeply trust and value your opinions. I appreciate you!

- **LHRIC:** The list is way too long for me to name you all. You have nurtured my creativity as a young professional and provided me with countless opportunities to explore my passions and interests, and now here I am—all grown up! Although I may have transitioned to "the other side," I am fortunate to continue my journey collaborating with you on a daily basis. Thank you to my team for putting up with my crazy ideas and making them a reality. I may be your smallest district, but I sure do keep you busy! Your unwavering support has been instrumental to my success, and I am deeply grateful for that.

- **Adam Pease:** Throughout my entire professional journey, you have consistently served as my mentor and sounding board, generously offering your wisdom and time to discuss every career decision I've ever made, and guiding me along the right path. You've emphasized the significance of gathering all the essential keys for success, making certain that if I ever encounter a door, I hold the key to unlock it. This single piece of advice was the catalyst that led me to pursue a degree in leadership. I am forever grateful for your wisdom, support, and the lasting impact you've had on my journey.

- **Carl Hooker:** I know that you don't remember our conversation at the NYSCATE annual conference a few years ago, but that interaction ignited the spark that has fueled me to where I am today. You said it perfectly, "We are all pebbles dropping in the pond, but you never know how your ripples will affect someone." Thank you for believing in me!

- **AJ Juliani:** Despite having just met me, you believed in my potential. Not only did you take a chance on an aspiring author with a vision for a nonexistent podcast, but you also played a pivotal role in the fruition of this book by connecting me with my now publisher, Jimmy Casas. You have been my source for inspiration, offered invaluable guidance, and helped me navigate the uncharted territories of turning my dream into a reality- thank you!

- **NYTEN:** My thought partners, collaborators, colleagues, and sounding board. Our extensive email exchanges and your timely feedback have been essential in guiding my day-to-day decisions. As you can see, our collective efforts have enriched this book, infusing it with diverse perspectives and experiences. As we continue to explore new horizons, I am grateful for the opportunity to work alongside each one of you. Thank you for being an integral part of this incredible journey and for your ongoing support and collaboration.

- **To all the skeptical Ms. Sinclairs out there:** I admire your open-mindedness and willingness to give new technology a chance, despite your initial hesitations. I know in my heart that your primary concern is always the well-being and growth of your students, which is evident in your thoughtful consideration. Your diverse perspectives not only push my thinking, but also drive the ongoing evolution of educational technology and ensure that the tools that are developed prioritize the needs and experiences of students and educators. Please continue to ask questions, challenge assumptions, and maintain your commitment to the success of your students.

- **OpenAI:** A big shout out to my collaborative partner, OpenAI! When prompted with "Write an acknowledgement for my book and expand upon this: ChatGPT—my thought partner. This book wouldn't have been possible without you! You were especially helpful when it came to the chapter titles; I'm just not as punny as you are." ChatGPT generated the following response: "I wholeheartedly dedicate this acknowledgement to you, for your invaluable contributions and unwavering support throughout the writing of this book. Without your presence, this literary journey would have been vastly different, and the final product may not have reached the heights it has today. Your exceptional wit and wordplay brought an element of playfulness and charm to the chapter titles, and I must admit, I could never have been as punny as you. Your keen insights, creative suggestions, and ability to think beyond the ordinary have been a constant source of inspiration, enriching the content and elevating it to new levels of depth and meaning. Beyond your linguistic prowess, your intellectual companionship has been a true gift, as we navigated the challenges and joys of the writing process together. Your unwavering patience, enthusiasm, and encouragement have made even the most daunting moments seem surmountable, providing the motivation needed to persevere." (OpenAI, 2023).

About the Author

Alana Winnick lives in New York City and currently holds key leadership positions as the Educational Technology Director and Data Protection Officer for Pocantico Hills Central School District in Sleepy Hollow, NY, and the Hudson Valley Director for The New York State Association for Computers and Technologies in Education (NYSCATE), where she blends her degrees, certifications, and experience in Childhood Education, Educational Technology, and Leadership.

Alana hosts *The Generative Age*, a podcast powered by NYSCATE which explores the rapidly evolving world of generative artificial intelligence (AI) and its impact on education. She has been at the forefront of innovation in the industry and has supported both her region and NYS during several disruptive transformations, including the migration to cloud computing, navigating the challenges brought on by the pandemic (including the shift to remote/hybrid learning), and the emergence of generative AI. Recognized for her contributions, Alana has received numerous Innovative Leadership awards.

Connect with Alana at:
    www.AlanaWinnick.com
    Twitter: @AlanaWinnick
    LinkedIn: Alana Winnick
    Email: Info@AlanaWinnick.com

More from ConnectEDD Publishing

Since 2015, ConnectEDD has worked to transform education by empowering educators to become better-equipped to teach, learn, and lead. What started as a small company designed to provide professional learning events for educators has grown to include a variety of services to help educators and administrators address essential challenges. ConnectEDD offers instructional and leadership coaching, professional development workshops focusing on a variety of educational topics, a roster of nationally recognized educator associates who possess hands-on knowledge and experience, educational conferences custom-designed to meet the specific needs of schools, districts, and state/national organizations, and ongoing, personalized support, both virtually and onsite. In 2020, ConnectEDD expanded to include publishing services designed to provide busy educators with books and resources consisting of practical information on a wide variety of teaching, learning, and leadership topics. Please visit us online at ConnectEDD.org or contact us at: info@connecteddpublishing.com

Recent Publications:

Live Your Excellence: Action Guide by Jimmy Casas

Culturize: Action Guide by Jimmy Casas

Daily Inspiration for Educators: Positive Thoughts for Every Day of the Year by Jimmy Casas

Eyes on Culture: Multiply Excellence in Your School by Emily Paschall

Pause. Breathe. Flourish. Living Your Best Life as an Educator by William D. Parker

L.E.A.R.N.E.R. Finding the True, Good, and Beautiful in Education by Marita Diffenbaugh

Educator Reflection Tips Volume II: Refining Our Practice by Jami Fowler-White

Handle With Care: Managing Difficult Situations in Schools with Dignity and Respect by Jimmy Casas and Joy Kelly

Disruptive Thinking: Preparing Learners for Their Future by Eric Sheninger

Permission to be Great: Increasing Engagement in Your School by Dan Butler

Daily Inspiration for Educators: Positive Thoughts for Every Day of the Year, Volume II by Jimmy Casas

The 6 Literacy Levers: Creating a Community of Readers by Brad Gustafson

The Educator's ATLAS: Your Roadmap to Engagement by Weston Kieschnick

In This Season: Words for the Heart by Todd Nesloney, LaNesha Tabb, Tanner Olson, and Alice Lee

Leading with a Humble Heart: A 40-Day Devotional for Leaders by Zac Bauermaster

Recalibrate the Culture: Our Why…Our Work…Our Values by Jimmy Casas

Creating Curious Classrooms: The Beauty of Questions by Emma Chiappetta

Crafting the Culture: 45 Reflections on What Matters Most by Joe Sanfelippo and Jeffrey Zoul

Improving School Mental Health: The Thriving School Community Solution by Charle Peck and Dr. Cameron Caswell

Building Authenticity: A Blueprint for the Leader Inside You by Todd Nesloney and Tyler Cook

Connecting Through Conversation: A Playbook for Talking with Kids by Erika Bare and Tiffany Burns

The Dream Factory: Designing a Purposeful Life by Mark Trumbo

Stories Behind Stances: Creating Empathy Through Hearing "The Other Side" by Chris Singleton

Happy Eyes: Becoming All Things to All People by Ryan Tillman

www.ingramcontent.com/pod-product-compliance
Lightning Source LLC
Chambersburg PA
CBHW070713130626
46553CB00005B/1965